T0231138

Electrophysical Phenomena in the Tribology of Polymers

Polymer Science and Engineering Monographs:
A State-of-the-Art Tutorial Series

A series edited by **Eli M. Pearce**, Polytechnic University, Brooklyn, New York

Associate Editors

Guennadii E. Zaikov, Russian Academy of Sciences, Moscow
Yasunori Nishijima, Kyoto University, Japan

Electrophysical Phenomena in the Tribology of Polymers

A. I. Sviridenok

Research Center on Resources Savings
National Academy of Sciences
Grodno, Belarus

A. F. Klimovich

Institute of Mechanics of Metal Polymer Systems
National Academy of Sciences
Gomel, Belarus

and

V. N. Kestelman

KVN International
King of Prussia, Pennsylvania
USA

Gordon and Breach Science Publishers

Australia • Canada • China • France • Germany • India •
Japan • Luxembourg • Malaysia • The Netherlands •
Russia • Singapore • Switzerland

Amsteldijk 166
1st Floor
1079 LH Amsterdam
The Netherlands

British Library Cataloguing in Publication Data

Sviridenok, A. I.
    Electrophysical phenomena in the tribology of polymers. –
    (Polymer science & engineering monographs ; v. 5 – ISSN 1023-7720)
    1. Tribology 2. Polymers – Surfaces 3. Polymers – Electric
    properties
    I. Title II. Klimovich, A. F. III. Kestelman, V. N.
    547.7′045′7

ISBN 90-5699-577-4

CONTENTS

INTRODUCTION TO THE SERIES

This series will provide, in the form of single-topic tutorial volumes, state-of-the-art information in specific research areas of basic applied polymer science. Volumes may incorporate a brief history of the subject, its theoretical foundations, a thorough review of current practice and results, the relationship to allied areas, and a bibliography. Books in the series will act as authoritative references for the specialist, acquaint the non-specialist with the state of science in an allied area and the opportunity for application to his own work, and offer the student a convenient, accessible review that brings together diffuse information on a subject.

PREFACE

Polymer materials come into contact with solids in the course of fabrication, processing and operation. Gradually, contacting surfaces disintegrate due to wear induced by friction.

The nature of physical and chemical processes governing and accompanying the tribodeformation and tribodisintegration of polymers is highly intricate and far from being well understood. Electrophysical phenomena in polymer materials, such as contact electrification and emergence of the electret state, are even less understood.

This area of polymer science has not yet reached the state of coherence. Numerous practical problems connected to triboelectric phenomena remain unsolved, both in fabrication and behavior of polymer threads and fabrics in a variety of industrial articles. In many respects, this is due to lack of knowledge on how contact electrification and electret effect cause friction and wear. In this book the authors strive to fill in gaps in the science of tribology, using results of unique studies.

The first experiments on electrification confirmed the presence of triboelectrification in all types of friction, for any combinations of contacting solid, liquid or gaseous bodies.

Modern ideas about surface states have allowed development of a physical model of electrification in solids and accumulation of more knowledge about the mechanism of triboelectrification in polymers. Studies of general properties of electrophysical, physical and mechanical processes and structural transformations of dispersed polymers under various mechanical influences are the most informative in this respect. Intensive dynamic contacts between dispersions are accompanied by deformations and structural transformations of polymers that raise local microcontact temperatures up to 10^3 K. A very strong electrification of dispersed polymers is observed. The electrical potential can reach 250 kV when confined to a closed cycle. A persistent electret state has been identified in mechanically activated polymers.

Natural and induced electret states in block polymers and composites strongly affect the parameters of friction and wear due to the structural transformations and superposition of electret state and triboelectrification. The current inversion, observed during friction between polar and non-polar dispersed block and composite polymer materials, is a general phenomenon in the process of triboelectrification.

Explanation of the mechanisms of such electrification and appearance of the electret state in friction of polymers uses basic postulates of the electrical theory of adhesion, some results of the electron theory of disordered systems and surface states. It has allowed us to identify links between processes of electrification and appearance of electrets at frictional contacts between polymer materials, and to outline available knowledge in the area of friction triboelectrophysics.

It should be emphasized that during our 30-year professional careers in studying the tribology of polymers, we have closely cooperated with many leading tribologists: K. C. Ludema, L. H. Lee and S. Jahanmir from the USA; D. Tabor, P. Jost, B. Briscoe and C. Taylor from the UK; R. Kurtel, D. Maugis and D. Barkens from France; K. Tanaka from Japan and others. Notwithstanding good personal and professional relations, there have always been some "specific" terminological barriers. These have been successfully removed through translations of publications and joint publications, notably: *Tribology in the USA and the Former Soviet Union; Studies and Applications*, eds. V. A. Belyi, K. C. Ludema and N. K. Myshkin (Allerton, 1994).

The major objectives of the present work are to inform English-speaking readers about achievements in the area of triboelectrical phenomena in polymers made by scientists from CIS countries and to standardize English and Russian terms, their compatibility and clarity.

This book is intended for researchers and industrial engineers; it can also be useful to graduate students and post-doctoral associates.

ELECTROPHYSICAL PHENOMENA IN THE TRIBOLOGY OF POLYMERS

A.I. Sviridenok, A.F. Klimovich, and V.N Kestelman

ABBREVIATION LIST

AE	auxiliary platinum electrode
DTA	differential thermal analysis
EEE	exoelectron emission
EIB	electron-ion bombardment
EIT	electron-ion technology
EMF	electromotive force
ES	electret state
ESCD	effective surface charge density
HDPE	high density polyethylene
HEE	high energy electronss
MCDP	momentary charge discharge pulses
Me	metal
PA	polyamide
PC	polycarbonate
PCA	polycaproamide
PE	polyethylene penton poly [3,3-bis-(chormetyl) oxetan]
PETP	polyethylene terephtalate (lavsan)
PFR	phenol formaldehyderesin
PMMA	polymetylmethacrylate
PP	polypropylene
PS	polystryrene
PTFCE	polyfluorine chlorine ethylene
PTFE	polytetrafluorethylene
PVB	polyvinylbutyral
PVC	polyvinylchloride
RCA	real contact area
RE	reference electrode
SC	semiconductor
SCC	short circuit current
SCD	space charge domain
SE	studied electrode
SS	surface
TGA	thermogravimetric analysis
TSC	thermally stimulated current
TSD	thermostimulated depolyarization
VS	volume states
WPC	wood-polymer composite

CHAPTER 1

PRINCIPLES OF FRICTION ELECTROPHYSICS IN POLYMERS

Accumulation and systematization of scientific information about the fundamental properties of friction have allowed formulation of the most important part of tribology – triboanalysis and its most essential branches, tribomechanics, tribochemistry, and tribophysics [1-3].

Tribomechanics deals with the surface contact problems, microscopic and submicroscopic processes in solids induced by tribomechanical effects and accompanying structural transformations. Tribochemistry analyzes the chemophysical and chemical changes created in solids by tribomechanical energy. Tribophysics deals with relations between frictional and physical effects at the interface between two solids or between a solid and its environment.

The tribomechanical energy alters the morphology, the electronic structure and the chemical composition of solids. Tribophysics identifies the leading atomic, microscopic and macroscopic physical processes evolving when solids are exposed to mechanical effects [4]: emission of photons; emission of electrons; static electrification; electrostatic discharges; emergence of the electret state; variations of the electrical conductivity; emission of lattice components; excitation of phonons; emergence and migration of lattice and electron defects; local heating in the solid phase; appearance of juvenile surfaces; surface expansion; cracking; abrasion; mass transfer between solids; amorphization; penetration of impurities; plastic deformation. They are all accompanied by complex dissipation processes due to different life-times of excited states and duration of relaxation (Table 1.1) [4].

Friction contact due to its discrete pattern is characterized by non-equilibrium processes and disintegration of a multitude of contacts with the parameters continuously varying in time and space. It is accompanied

3

Table 1.1. Duration of relaxation of different excited states induced by mechanical effects in solids

Cause of excitation	Time of relaxation, s
Impact process	10^{-6} (Hertzian impact time)
Triboplasma	1^{0-7}
Gas discharge	10^{-7}
Hot spots	$10^{-3}...10^{-4}$
Electrification	$10^{2}...10^{5}$
Emission of exoelectrons	$10^{-6}...10^{5}$
Triboluminiscence	$10^{-7}...10^{3}$
	(fluorescence, phosphorescence)
Lattice defects (e.g., V_k-centers in LiF at various temperatures	$10^{-7}...10^{6}$
Phonons	$10^{-6}...10^{-10}$
Cracking	$10...10^{3}$ cm/s (rate of disintegration)
Appearance of juvenile surfaces	$1...100$ ($1.3 \cdot 10^{-4}$ Pa); 10^{-6} (10^{5} Pa)
Life time of excited metastable states	$10^{-3}...10^{-2}$

by various electrophysical phenomena, such as static electrification, emission processes, emergence of the electret state, and variations of the electrical conductivity.

A large variety of non-correlated data related to these problems have been reported recently. Specifically, the electrification effect has been observed for all types of friction, phase and aggregate states of contacting bodies. Electrification with its post-effects is especially noticeable in polymers [5–10]. A relation has been reported [11] between high local temperatures and structural transformations of polymers, the rate of contact electrification and the appearance of the electret effect. On the one hand, when the temperature increases, the rate of electrification accelerates due to the growing density of surface states and the activity of charge carriers; on the other hand, the charge reduces due to the growing volumetric and surface conductivity of a polymer.

Friction interactions produce changes in the surface electrical parameters, the most important of them being the appearance of surface charge and the emission of electrons caused by the charge field. The emission of exoelectrons with the energy under 1 eV and electrons with a greater energies (up to 100 keV) [12,13] was measured. The energy spectrum and the rate of electron emission are governed by the dielectric behavior of a material, the nature and concentration of adhesion-active functional groups in a polymer, along with the external conditions. Interesting reports have appeared about the mechanochemistry of high molecular compounds and polymers in friction [14,15]. Experimental results have been reported [16] that show the similarity in variations of the extent of electrification, as soon as mechanodestruction of the polymers is intensified by friction.

Friction of solids including polymers produces acoustic vibrations and such vibrations affect the friction process. Relations between the intensity of acoustic emission, the duration of operation and its load, as well as the environmental characteristics, the wear pattern, and other factors, have been investigated [10,17–19].

The experimental and theoretical results show a significant role of electrical phenomena caused by friction in the processes of hydrogenation of metals [20–22], tribopolymerization [23,24], frictional transfer [5,21,25–27], including the sign-variable transfer [28]. The analytical review of a number of reports [21,29–31] indicates that static discharges initiate wear of contacting bodies which causes electrical erosion. It should also be mentioned that the most probable mechanism of the electrical destruction of solid organic dielectrics is the treeing promoted by repeated partial discharges, moisture, and so forth.

Hence, there is a need to make a systematic review and the analysis of the electrophysical triboeffects mentioned above.

1.1. ATOMIC AND MOLECULAR ELECTROPHYSICAL CONCEPTS OF THE NATURE OF FRICTION IN POLYMERS

The early theories of friction in solids were based on mechanical models. With the progress of atomic and molecular ideas, the novel theories evaluating adhesion in the frictional contact zone were put forward [5,32–34]. The molecular theory of Deryagin [35] had become a promising tool for evaluating the electrophysical phenomena in friction through studying the properties of friction forces and adhesion based on the statistics of coupled interactions between monocrystals.

The molecular friction theory yields a binomial expression for the external friction force F as a function of the nominal load N:

$$F = \mu N + \mu p_0 S_0 \quad \text{or} \quad F = \mu (N + p_0 S_0), \qquad (1.1)$$

where μ is the true friction coefficient; S_0 is the real contact area; p_0 is the specific adhesion affecting area S_0.

It follows from expression (1.1) that the force of friction is the sum of two components. One of them, μN, depends on the response of Born forces to the load, the other, $\mu p_0 S_0$, depends on the response of repelling forces to the force of molecular attraction.

The adhesion theory of Deryagin [36], describing interactions between two contacting solids is based on the Helmholtz's idea of a double electrical layer [37] and his unique experimental and theoretical studies. Therefore, this theory was termed as the electrical one. The adhesion may also be caused by the Van der Waals forces or by the appearance of chemical bonds between atoms along the interphase boundary. The electron adhesion theory [38] emerged from the electrical theory with the evolution of the ideas about surfaces of solids with active electrons and with the development of the theory of surface states, which was rated in [21] as being "the one with a greater heuristic potential with respect to friction physics." This theory has been corroborated by recent studies that show that the investigation of electrophysical friction phenomena is one of the most valid and most effective ways of properly studying the nature of friction. Although, the above friction theories ignore the force of non-contact adhesion. This fact resulted from the absence at that time of the studies on mechanism of discrete contacts in solids [39].

It is essential to note that the state-of-the-art computations in the area of the friction theory have been evolving together with the development of fundamental electromagnetic theory of interactions in condensed phase. They in turn promoted the development of both the electrical and, later, the electron theories of adhesion between solids. Based on the knowledge of emission and absorption characteristics of electromagnetic waves produced by contacting bodies, the electromagnetic theory of molecular forces [40,41] was developed. It resulted in the theory of selective resonance mechanism of energy dissipation for external friction and in the development of the oscillatory model of friction in solids by Bufeev [42], as well as the friction theory by Postnikov [21].

Discovery of the special role of the donor-acceptor effects in the electron friction theory and application of the phenomenological approach to the description of the donor-acceptor bonds using statistical evaluation of the appearance of donor-acceptor bonds have obviated intricate, quantum mechanical computations for determination of the wave functions of heteropolar molecules and the energy of a multielectron system. According to these models, the charges producing a double electrical layer resulting from redistribution of the electron gas density in the contacting bodies' interface, do not occupy any definite volume. Instead, they have a planar distribution with a constant density σ.

Hence, the force of interaction between two unlike contacting bodies can be determined as the attraction force between the unit area of the capacitor plates :

$$f_e = 2\pi\sigma^2. \qquad (1.2)$$

For simplicity, expression (1.2) ignores the average dielectric permeability.

The parallel plate capacitor model uses number of assumptions. The true structure of the double-electrical layer, specifically, in contacts between metals and semiconductors, metals and dielectrics, is more complex. Yet, assuming the surfaces to be homogeneous and having typical discrete structure, the electron theory derives an expression for the specific adhesion force:

$$f_e = e \sum_{i,k} E_z(X_i, Y_k), \qquad (1.3)$$

where X_i, Y_k are the charge coordinates in one of the charged planes; E_z (X_i, X_k) is the field produced by the charges of the other plane at the point where the charge is located.

The same field affects the charge at any point (X_i, Y_k), therefore

$$f_e = \sigma E_z(X_i, Y_k). \qquad (1.4)$$

If the charges on the plane are positioned at the vertias of the lattice with the constants a_1 and a_2, and the planes are separated by distance z, then, for $z \gg a_1 \sim a_2$ the true field $E_z(X_i Y_k)$ representing a periodic function with periods a_1 and a_2 approaches a one-row field of a uniformly charged plane, $\langle E_z \rangle = 2\pi\sigma$. For $z \ll a_1 \sim a_2$ the effect of all the charges of the opposing lattice on any charge, excepting the closest one, can be ignored, so that

$$f_e = \frac{\sigma e}{z^2} = \frac{ne^2}{z^2}. \qquad (1.5)$$

Here n is the number of charge pairs per unit area.

When the parallel plate capacitor model is applied, these assumptions yield a reduced value of f_e, therefore substitution of the true field with the average field of uniformly distributed charges is justified, provided that z is much greater than the average spacing between the charges. Calculations have shown [38] that the adhesion forces can be estimated in a quantitatively correct way for z ~ 10...100 Å and larger. Such values of thickness of the double electrical layer is typical for contacts between semiconductors. Therefore, when the adhesion electrostatic component is studied, the primary case is the metal-semiconductor contact with any zone structure and spectrum of the surface states. Assuming that a microscopic gap d with the dielectric permeability ε_3 separates the semiconductor from the metal, an expression has been derived to determine the field in the gap, E_3, and the surface potential, φ_t:

$$E_3 = \frac{\varepsilon_t KT}{\varepsilon_3 e L_D} \sqrt{\frac{2}{e n_t}} \int_0^{\tilde{\varphi}_n} \sigma(\tilde{\varphi}_n) d\tilde{\varphi}_n + \frac{4\pi}{\varepsilon_3}(\tilde{\varphi}_n), \qquad (1.6)$$

$$V_k^0 = \tilde{\varphi}_n + d\left[\frac{\varepsilon_t}{\varepsilon_3 L_D} \sqrt{\frac{2}{e n_t}} \int_0^{\tilde{\varphi}_n} \sigma(\tilde{\varphi}_n) d\tilde{\varphi}_n + \frac{4\pi}{\varepsilon_3}(\tilde{\varphi}_n) \right]. \qquad (1.7)$$

Here $\tilde{\varphi}_t = -\dfrac{e\,\varphi_t}{kT}$ is the dimensionless electron potential energy; $\sigma(\varphi_t)$ is the surface charge; n_i is the average concentration of carriers typical for a given semiconductor; ε_t is the semiconductor dielectric permeability; $L_D = \sqrt{\dfrac{4\pi e^2 n_t}{\varepsilon_t kT}}$ is the Debye's screening length; $V_k^0 = \varphi_t + dE_3$ is the potential difference between the metal and the semiconductor with no surface charges.

Using equations (1.6) and (1.7) and knowing the zone structure and the spectrum of semiconductor surface levels the field strength, the gap and the adhesion electrostatic force can be determined:

$$f_e = \varepsilon_3 E_3^2 / 8\pi. \qquad (1.8)$$

In particular, using the derived expressions and considering just the two types of surface levels (the donor and the acceptor), it has been shown that the field in the gap is $E_3 \approx 10^9$ V/m, when the surface state density is $10^{14}...10^{15}$ cm^{-2}, whereas the adhesion specific force f_e reaches 4 MPa.

The results testify that the electrical fields in the metal-semiconductor interface and between metals may be of the same order of magnitude; also, stronger fields and adhesion forces in the gap are reached when a semiconductor has a greater number of surface states.

Other theoretical and practical aspects relating to the role of double electrical layers in contacts between solids are also worth considering. In 1952 Karasev, Krotova and Deryagin discovered the emission of fast electrons when the contact of solids is broken in the vacuum. This discovery stimulated a variety of studies of the emission phenomena.

Analysis of the exoelectron emission [43] has revealed that the charge on the inner side of the polymeric film delaminating from a solid substrate generates the field sufficient to produce emission of high energy electrons from the film outer surface [44]. Also, in the process of cohesive disintegration of dielectric materials or adhesive bonds, the surfaces of dielectrics behave differently from common surfaces. These differences are due, firstly, to the surface juvenility, *ie* to the adsorptional purity of the appearing surfaces, and secondly, to their specific energy state. Disintegration of various dielectrics in the vacuum is accompanied by the emission of electrons with the energy up to 10^2 keV, and various fragments of the disintegrated dielectrics have a "mosaic" charge pattern instead of the unipolar one.

It has been noted that friction can produce an autoelectron emission from the surface states of the charged dielectric. Exoelectrons accelerating in the surface charge field can contribute to the high energy emission .

It was discovered [45] that disintegration of the adhesive contact between two solids and machining of some materials can cause photo emission. Detected photon flux had the rate of $N_p = (1...5) \times 10^3$ pulses per second. As a rule, such emission occurred only at the moment when adhesion bonds failed and produced significant electron emission. Emission of photons stops once the adhesive contact fully disintegrates, whereas emission of electrons can persist for a long time, sometimes several hours [46].

These results show that the emission of photons, when the adhesive contact is lost, results from deceleration of high energy electrons in the substrate and can be characterized as X-rays. A team of researchers headed by Deryagin registered the discovery on the June 7, 1984 stating that "...Prior to destruction of materials, detectable X-ray emission appears in originating cracks..." (Note that until 1991, in addition to inventions, scientific discoveries were also registered in the USSR .)

Other features of contacting systems related to the governing role and behavior of the double electrical layer should be noted. Primarily, the electrostatic component of the adhesion force, *ie* the attractive force between the double layer planes, has significantly longer range of action than the molecular forces and it recedes more strongly in response to time than to distance.

The electron adhesion theory for polymers can be useful in explaining the electrical nature of adhesive-cohesive compounds, as well as the effects of velocity, pressure, and the nature of the surrounding atmosphere on processes of gas discharge, adhesion and seizure. It should also be remembered that the friction is affected by the adhesion forces over the real contact area and the non-contacting sites on the surfaces within the radius of mutual molecular attraction [39].

Now let us consider the electromagnetic theory of molecular forces based on the ideas of Lebedev about emission and absorption of electromagnetic waves by contacting systems of atoms [47]. These ideas were further developed by Lifshits [48] who conceived that a condensed body is a source of a fluctuating electromagnetic field. The theory treats the fluctuations of electromagnetic fields emitted by representing the mutual action of the two condensed phases as an interaction of two giant molecules - as in the case of the donor-acceptor mechanism. It allows estimation of the force of mutual attraction between the bodies conditioned by Van der Waals forces responsible for the non-contact adhesion as an electromagnetic component of adhesion force f_m.

If the spacing H between the bodies exceeds the major wavelengths of the spectrum of emission (absorption), and the temperature approaches zero, providing that f_m depends only on the dielectric permeability in the stationary field, the formula to determine f_m becomes [48]

$$f_m = \frac{\pi}{240} \cdot \frac{hc}{H^4} \left[\frac{\varepsilon(0) - 1}{\varepsilon(0) + 1} \right]^2 \psi[\varepsilon(0)]. \qquad (1.9)$$

Here $\varepsilon(0)$ is the dielectric permeability in the stationary (electrostatic) field; the function $\psi[\varepsilon(0)]$ can be specified by plots produced by numeric integration.

In this case $\varepsilon(0) = \infty$, $\psi[\varepsilon(0)] = 1$ for metals and the electromagnetic component of the adhesion force is

$$f_m = \frac{\pi^2}{240} \cdot \frac{\eta c}{H^4} = \frac{0.013}{H^4}. \tag{1.10}$$

When H values are small compared with the major absorption wavelengths, the force of attraction by condensed phases is proportional to H^{-3}. Using mica specimens, it was demonstrated that transition from the law of interaction between bodies $f_m = cH^{-3}$ to the law $f_m = cH^{-4}$ occurs at $H = 100...200$ Å.

Theoretical results obtained by Lifshits in the area of electromagnetic theory of molecular forces were experimentally confirmed [40]. A unique technique for measuring molecular forces using analytical scales with the feedback has allowed direct measurement of the interaction force F between flat and convex (the curvature radius 26 cm) quartz plates spaced at $H = 0.1...1$ µm and to obtain the experimental dependence $F = f(H)$ that agrees well with the theoretical curve $F = cH^{-4}$ obtained by Lifshits. The theory was later applied to describe molecular interactions in metal-polymer contacts [49].

Postnikov [21] has demonstrated that two adhesion (electrostatic and electromagnetic) components analyzed above should be taken into account as the components of the total attraction force in the binomial friction law.

Taking into account the non-contact adhesion forces between the surface sites related to the contact area (S_n-S_0), expression (1.4) can be rewritten as

$$F = \mu[N + p_0 S_0 + p_m (S_n - S_0)], \tag{1.11}$$

where p_m is the average specific attraction force between condensed phases; S_n is the effective contact area.

When the donor-acceptor pattern covers the entire contact area, the specific pressure p_0 in formula (1.11) has the physical sense coinciding with that of the electrostatic adhesion component f_0. The magnitude of the specific attraction force p_m is the same as that one of electromagnetic adhesion component f_m.

Dubinin [29] has considered friction surfaces as plates of the parallel plate capacitor generating the electrical charges. He has shown that friction processes result in the high amplitudes' potential variations whose frequencies are small for the surface layer microparticles and high for submicron ones. He concluded after some experimentation that the degree of excitation of atoms and the atomic lattice vary by friction throughout the depth. It is maximum for the submicroscopic rough profile of the surface layer and gradually reduces further from the surface. Thus, the mechanical energy during friction of solids transforms into the oscillatory and undulatory energy of submicroscopic and microscopic profile and structural components of the surface layer. This, in turn, leads to the appearance of electrical, thermal, acoustic, and other phenomena characterizing friction in a qualitative manner.

Application of the electromagnetic theory of molecular forces to the friction couples with one or two polymer members having strong selectivity in the electromagnetic radiation absorption, has established a connection between the electromagnetic component of the adhesion force and the dielectric properties of contacting bodies and the environment. This helped to explain the mechanism of polytetrafluorethylene self-lubrication [49,50].

The theories based on oscillatory and resonance-selective friction models are examples of successful application of the electrophysical approach to the nature of friction. The oscillatory model assumes that the energy dissipation at external friction is primarily due to the electromagnetic interaction between condensed phases.

Postnikov [21] has considered the process of constrained oscillations of identical oscillators in the rubbing bodies as an elementary friction mechanism. In this case, positively charged atomic fragments undergoing harmonic oscillations with frequency ω_0 act as identical oscillators. Thus, he derived an expression for the friction force:

$$F = \frac{\delta q^4 S}{8\pi\varepsilon_0^2\, ma^3 Q(Z_0 - bN)^4\, \sqrt{(\omega_0^2 - \omega^2)^2 + 4\beta^2\omega^2}}. \qquad (1.12)$$

For the friction model of relaxation type (at $\omega \ll \omega_0$), one gets

$$F = \frac{\delta q^4 S}{8\pi\varepsilon_0^2\omega_0^2 ma^3 Q(Z_0 - bN)^4}, \qquad (1.13)$$

where δ is the factor depending on the type of mating lattices; m and q are the ion mass and charge; S is the contact area; a is the lattice constant or the period of identical arrangement of oscillators; Q is the experimentally estimated quality factor of external oscillators ; b is a constant; N is the "normal load".

In a more general case, resolving the external force in a Fourier series

$$f(t) = \sum qE_{\alpha k} \cos(\omega_{kt} - \alpha_k) \qquad (1.14)$$

and considering that several oscillators (including linear ion chains oriented along the Z axis) with the parameters q_i, m_i, ω_{0i}, β_i, Q_i, may appear on rubbing surfaces, the following expression for the force of friction can be derived:

$$F = \frac{S}{8\pi\varepsilon_0^2} \sum_i \sum_k \frac{n_i q_i^4}{m_i a_i Q_i(k)(Z_{0i} - bN)^4 \sqrt{(\omega_{0i}^2 - \omega_k^2)^2 + 4\beta_i^2 \omega_k^2}}. \qquad (1.15)$$

Here $i(k)$ indicates the functional relation between Q_i and ω_k.

The analysis of the derived expression indicates that the response of the friction forces to the velocity $F(v)$ contains a number of alternating maxima and minima produced by the resonance-selective mechanism of energy dissipation at friction. The friction forces reach the maximum values at the dynamic resonance of one of the major groups of equivalent oscillators, ie at velocities when

$$\omega_k = \sqrt{\omega_{0i}^2 - 2\beta_i^2}. \qquad (1.16)$$

Using these expressions to analyze functional relation between the various parameters, for example, F and Q or ΔT and Q (ΔT is the temperature increment due to friction), we have concluded that the external friction mechanism in real materials had much in common with the internal friction mechanism of the hysteresis type. Then, knowing, for example, the area of the dynamic hysteresis loops, it is possible to estimate the friction force. Yet, resonance, hysteresis, and relaxation coexisting in external friction processes contribute differently to the intensity of energy absorption and, hence, to the friction process mechanism.

The relaxation model of friction and wear in polymer materials [51,52] being a generalization of the molecular kinetic theory and the theory of physical nodes presents a certain interest. This model contains the following analytical expression for the interphase interaction energy:

$$E(t) = N_a^0 \varphi(t)\overline{U}(t) + \int_0^t \upsilon_a(\tau)\varphi(t-\tau)\overline{U}(t-\tau)d\tau. \qquad (1.17)$$

Formula (1.17) allows consideration of the features of contact processes in polymer surface layers.

Development of the relaxation model for the dynamic contact between polymers and metals has derived the following expression for the adhesion component of the friction force [51]:

$$F_a = \frac{1-b}{v_c}\left[N_a^0 \frac{d\varphi(t)}{dt}\overline{U}(t) + \int_0^t \upsilon_a(\tau)\frac{d\varphi(t-\tau)}{dt}\overline{U}(t-\tau)dt \right]. \qquad (1.18)$$

Taking into account the relations reported in [52] and the basic ideas of the thermal fluctuations theory of polymer disintegration [53], the following expression has been derived for the cohesion component of the friction force:

$$F_k = bW_2 N_a \frac{(D+\chi_{II})}{\lambda v_c}\left(U_k + kT\ ln\frac{v_c \tau_{g.0}}{D+\chi_{II}} \right). \qquad (1.19)$$

For initial friction stages ($t < t_1$, $t < t_2$) the expression of the friction force has the form:

$$F_{fr} = \frac{W_2}{v_c}\Big\{ (1-b)U_0 N_a^0[1-(bW_2+B_c)t] +$$

$$+bN_a^0\frac{1}{\lambda}(D+\chi_{II})(1-bW_f)\Big(V_k + kT\ ln\frac{v_c \tau_{g.0}}{D+\chi_{II}}\Big)\Big\}. \qquad (1.20)$$

In late stages of friction, the effect of structural modification on F_{fr} becomes more complex: the function F_{fr} may have one or several extrema illustrating the competing effects of molecular mobility, restructurization, and wear on the friction force. Electrophysical phenomena, the formation of the electret state, in particular, make their own contribution in the friction of polymers.

Thus, by studying the friction contact electric states for polymers rubbing against metals and dielectrics (recent results), it has been established that   the local electrical fields produced by electrostatic charging in the contact zone significantly affect the physical and chemical behavior of contacting surfaces, as well as friction and   the wear of polymers [11,21,26]. The electrification by friction is primarily explained by the contact electrification caused by the formation and disintegration of the double electrical layer at the moments of the establishing and failure of the contacts [38].

Notwithstanding extensive studies, the role of electrification processes in the friction of polymers remains ambiguous.

1.2. SURFACE STATES. THE ELECTRON THEORY OF DISORDERED SYSTEMS

Let's consider some theoretical aspects of the fundamentals of polymer electrophysics. The surface electron states are the centers of localization of free charge carriers playing a crucial role in the processes occurring on the surfaces of semiconductors and dielectrics. Their existence was were theoretically predicted by Tamm [54], Shockley [55] and Bardeen [56] and experimentally discovered by Shockley and Pearson [57].

Unlike "theoretical" solids, the real ones always have some surface defects, *ie* local discontinuities of a strictly periodical surface structure. One distinguishes the "biographical" defects defined by the history of the surface machining and treatment,   as well as thermal and adsorption defects [58]. Because the "biographical" defects represent a random disordered   fraction of atoms on the surface, as well as   dislocations, vacancies, domains, steps, and so forth, a new system of electron surface states emerges in the "forbidden" zone. The contact of the surface with the environment results in adsorption of atoms and molecules and appearance of adsorption   states   superimposing   on   the   energy   spectrum   of "biographical " surface states.

Defects may be influenced by physical and chemical properties. Surface tribophysical properties are strongly influenced by the physical defects. This is demonstrated particularly by the studies of trapping mechanisms using the techniques of thermally- and photographically-stimulated conductivity [59,60]. These studies showed that the parameters of traps are determined by physical rather than   chemical defects.

The situation is more complex in polymers than in metals due to a much greater concentration of defects and   the presence of multiple intermolecular defects (kinks). For example, their concentration in polyethylene after tenfold extension may reach 10^{21} cm^{-2}. These defects fluctuate and can migrate in the volume emerging on the surface [61]. Moreover, wider spectra of relaxation time $(10^{-10}...10^{10}$ s) typical for polymers evidence a high mobility of various structural components.

All types of defects are responsible for the appearance of an additional system of electron surface states in the "forbidden zone" that alter the spectrum of the eigenstates. "Biographical," thermal, and adsorption surface states are usually distinguished. In addition, the terms "active centers," "active surface spots," "active elements of solids," "active elements of real structures," "electrically active surface spots," "fast states," "slow states," "volume states," "capture levels," "traps," "recombination centers," "capture centers," and others [38,61,62,63,64] are also used.

Extensive studies lead to the model of real surface (Fig. 1.1.) stating that the energy spectrum of the semiconductor, oxide heterotransition includes three groups of surface states: (*1*) fast states in the direct contact with the semiconductor; (*2*) slow states within the oxide subsurface region of $\approx 10...20$ Å thick, exchanging charge carriers with the semiconductor according to the tunneling mechanism; (*3*) superslow states of the oxide with the over-barrier mechanism of interactions with the semiconductor (without any sharp borderline between the regions occupied by slow and superslow states).

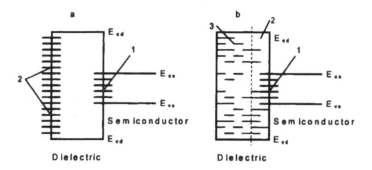

Figure 1.1.   Diagram of real surface zones [58]: 1, fast surface state; 2, slowsurface state; 3, superslow surface states. E_{cs}, E_{cd} – conductive zone boundaries; E_{vs}, E_{td} – valency zones of semiconductor and dielectric, respectively; a — slow states arranged on semiconductor oxide film external surface; b — slow states localized along dielectric–semiconductor boundary.

Due to the existence of surface states   full of charge carriers, an electrical charge appears on the surfaces of semiconductors and dielectrics. The full surface charge in general is the sum of charges in fast and slow surface states [58,65]:

$$Q_s = Q_{sf} + Q_{ss}.$$ (1.21)

Then, for fast processes of electrification due to   friction when the contact lasts 10^{-5} s, the full surface charge of the polymer is   the sum of charges in fast and slow states. In   relaxation processes related to the electret effect in polymer materials   the total charge is primarily determined by the charge in slow surface states, *ie* by Q_{ss} [66].

Ionization of surface states near the boundary of even an insulated semiconductor (dielectric) produces a double electrical layer with the density of charges and the structure   mostly determined   by the presence and the characteristics of the surface states.

The surface states   very strongly affect the structure of the double layer appearing in the   contacts: semiconductor (dielectric) – semiconductor (dielectric) or semiconductor (dielectric) – metal. In this case high density of surface states induces the effect of surface "metallization" when the charge of the double layer is produced by the ionization of surface states. The   order of magnitude   of this charge is comparable with the charges of the layers appearing in the contact between metals [12].

Krupp [67] has investigated the contact between very pure metals and has shown that the experimental value of surface density of charges, 10^{10} electrons/cm^2 (10^{-9} C/cm^2), approximately corresponds to the density of surface states $D_s = 10^{-10}$ cm^{-2}. This points out the correlation between the density of surface states, the concentration of structural defects, and the surface density of charges.

Studies of the contacts between various metals and atactic polystyrene (PS) have revealed that the volume charge characteristics of the latter compound at points where it contacts a metal are defined by the surface states with the density 10^{16}/cm^2 and the volume state with the density 10^{14}/cm^3 distributed within the boundary layer of about 4 μm (possibly up to 15 μm) thick [68].

The density of the surface states in the PS metal contact also depends on the type of a metal [68]:

Metal	Density of surface states n_s, cm^{-2}
Pb	$1.7 \cdot 10^{10}$
Au	$3.0 \cdot 10^{10}$
Ni	$1.3 \cdot 10^{10}$
Pt	$7.0 \cdot 10^{9}$
Sn	$2.0 \cdot 10^{9}$

The metal substrates in metal-polymer compounds favor higher concentrations and deeper penetration of carriers of the electron nature. Hence, Fabish [68] considers two models to investigate the processes of transfer and exchanges of charges in the metal-polymer contact. In the first model, the dielectric is represented as a semiconductor with a wide forbidden zone [85]. In the second model, only the surface states are considered [67,70].

It turned out that the charge injected into the polymer volume and surface states would reinforce the electric field between the dielectric and the metal (this field governs the discharge processes).The energy of the volume and the surface states depends on the film thickness. When the contact is disrupted, significant losses take place only on the surface levels due to the poor mobility of carriers (10^{-11} cm^2/s) on internal levels. Calculations have revealed that the share of the volume charge is 2/3 (the PS – Pb contact) and 1/3 to 2/3 in the contact of PS with indium. Metals with the small work function interact, as a rule, with the volume (V) states, whereas metals with the high work function interact with S-states. Charge carriers in the metal-polymer contact are electrons; the injected charge is composed of electrons transferred from the metal into the traps in the polymer. The metal-polymer contact system is always in the non-equilibrium thermodynamic state.

Cessler also demonstrated that the surface and the volume charges in electrets are localized at the capture levels with the energies within the forbidden zone separating the conductivity and the valency zones [69]. Two kinds of traps are distinguished: the electron trap and the hole traps. The electron traps are neutral in their free state and negatively charged at ionization. On the contrary, the hole traps are neutral in the "filled" states and are charged positively when electrons are released. Both the valency and the conductivity zones are distributed throughout the volume of dielectrics with the periodic lattice structure and sufficient overlapping of the orbital states of adjacent molecules. Such materials have a set of discrete capture levels.

Local energy levels in amorphous, polycrystalline, or partially crystalline substances to which polymers belong are formed under the influence of the neighboring molecules. Appearing zone structures are separated by potential barriers. As a result, each atom or group of atoms acquire a set of their own energy levels (Fig. 1.2).

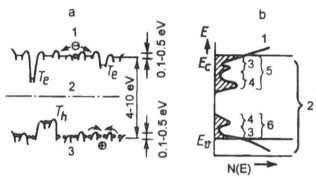

Figure 1.2.   Polymer energy diagram (a): 1, conductivity zone; 2, forbidden zone; 3, valence zone (T_e — electron traps; T_h — hole traps); density of polymer states N(E) (b): 1, free; 2, localized states; 3, shallow; 4, deep; 5, electron; 6, hole traps (overshadowed are the localized states-traps; E_c and E_v — mobility edges) [70].

Such materials have discrete capture levels or bands of capture levels. Particularly, from two to six discrete capture levels released within the temperature range 293...473 K were revealed for Teflon Polyfluorethylenepropylen.

Figure 1.2 shows one possible distribution of the density of states. Instead of the boundaries of the conductivity and the valency zones, one can observe mobility edges E_c and E_v along which the mobility of the charge carriers changes sharply.

Above it has been pointed out that the emergence of surface and volume states in polymers is due to a number of structural abnormalities – defects of monomer units, irregularities in chains, imperfections of crystallites, the presence of a large number of interphase boundaries, impurities, and so forth.

Bauser [70] had studied the volume state in polyethylene (PE) and showed that the depth of a hole trap equaled the difference between the energies of ionization of an isolated PE-molecule and a molecule with capture capabilities. The capture depth of an electron trap equals the corresponding difference between the electron affinities. Calculations have revealed that the structural defects, like carbonyl groups or double bonds, yield shallow or intermediate capture levels at depths of up to 1 eV. Clustering of heterogeneous molecules produces deeper capture levels. The series of the capture levels with the depth of about 0.4 eV correspond to the presence of methyl groups.

Chemical and physico-chemical transformations and ionization of polymers due to friction favor higher density of surface and volume states. There are several possible ways to defect ionization for ion crystals depending on the level origin. These include spontaneous or thermal filling of the defect levels by electrons from the valence zone, charge redistribution, electron transferring from charged impurity to defects.

Ionization and migration of charged defects, as well as polar groups orientation produce electrical fields on the surface and as well as within the volume. Furthermore, relaxation of elastic stresses is often closely tied to the relaxation of electrical fields. Its external macroscopic manifestations are changes of electrical conductivity, gas discharge, recombination luminescence, emission processes, and so forth Butyagin [72] asserts that "...deformation of interatomic bonds and structural decomposition of a solid caused by mechanical forces, the crystal electron subsystem disturbances with ionization of defects and appearance of electrical fields should be considered as a different aspects of the same phenomenon. Its nature does not depend on the way the internal stresses were created. In each specific case, such as cleaving or cutting, friction ... specific channels of elastic energy relaxation would dominate."

Seggern [71] studied corona electrets of Teflon Polyfluorethylenepropylen and successfully distinguished the surface states from the volume states. He determined typical temperatures at which traps started to release charges. They were 423 and 473 K for surface and volume states, respectively, and 443 K for the "intermediate" trap localized near the surface. Moreover, another tiny capture level was detected in the region oaround 368 K (Table 1.2).

According to the data of Cessler [69], Table 1.3 lists the maximum densities of accumulated charges and densities of volume traps for dielectric films. The maximum charge densities correspond to the filling of both the volume and surface trapping levels. Two circumstances should be noted. Analysis indicates that in some polymers, for instance, in Teflon, the surface electron states dominate. Because the electrification process is restricted by the "sparking" conditions (for the case of the corona electrification) and by other experimental conditions, the surface charge density turns out to be less than the value corresponding to the maximum possible filling of the traps.

Table 1.2. Characteristics of traps distribution for negative charges in Teflon Polyfluorethylenepropylen film 25 μm deep.

Peak temperature, K	Position with respect to charged surface, μm	Trap type
368	0...25	Shallow (based on energy), active after complete filling
428	0...0.5	Surface
443	0.5...1.8	Near the surface
473	1.8...25	Volume

Investigation of the space distribution of electret charges in various polymers has shown the effect of polymer properties and the specimens thickness, as well as electrification conditions.

Table 1.3. Maximum observable charge densities and densities of completely filled traps in unilaterally metallized dielectric films [69].

Material	Thickness, μm	Surface (S) or volume (V) charge	Projected charge density* with its sign indication 10^{-8} C/cm^2	Density of completely filled volume traps (sign is shown in parenthesis) 10^{15} cm^{-3}
Polyfluorethylenepropylene	12.5	Mainly S	0.5 (+,-)	
Polyfluorethylenepropylene	25.0	V		0.14** (-)
Polyethylenephtalate	3.8	S+V	1.2 (+,-)	
Polyethylenephtalate	4...6	V		10 (?)
Polyacrylonitrile	2.0	S+V	1.0 (+,-)	
SiO$_2$	0.06...0.1	S+V	4 (-), 2 (+)	

*   Effective surface charge density [73]
** Attributed to deep traps with temperatures of relaxation of thermally stimulated current being 479 K and higher.

To conclude, it should be mentioned that the internal relaxation of the charges in dielectrics (including polymer electrets) is governed by the conductivity phenomena which, in turn, are defined by the above characteristics, such mobility of carriers, their concentration, conditions of carrier injection on electrodes, and so forth.

The electron theory of disordered systems along with the theory of surface states are the effective tools for analysis of the electrophysical phenomena of polymer friction.

1.3. STATIC TRIBOELECTRIFICATION AND ELECTRET EFFECT IN POLYMERS

The classic concept states that static electrification covers the processes leading to the appearance and separation of positive and negative electrical charges produced by mechanical deformations at collisions or static contacts between two solids, or between solids and liquids, as well as at separation of surfaces. Contact electrification, triboelectricity of block specimens, as well as electrification of powders should also be mentioned here [74].

Lennard [75] and Coehn [76] pioneered the studies of static electrification. Findings of the Russian scientist Gezekhus [77] from the same period showed that the friction of chemically-identical solids with different densities leads to positive electrification of the one with the higher density.

As early as 1757, Wilke observed that some substances, such as amber, glass, wool, silk, could be arranged in triboelectrical series. Coehn pointed out the connection between the position of a dielectric in the triboelectrical series and its electrical characteristics. He showed that at the contact of two dielectrics, the one with greater dielectric constant acquired positive charge. The charge of the two contacting dielectrics is proportional to the difference between their dielectric constants. For the dielectric-metal contacts, metal can acquire both positive and negative charge.

Using Coehn's rule, Richards derived the relation

$$Q = C(K_1 - K_2),   (1.22)$$

where Q is the trieboelectrical   quantity; K_1, K_2 are the   dielectric permeabilities; and C is the coefficient of proportionality.

During the last decade, many researchers have tried to answer the question about the nature of static electricity. They investigated the mechanism of static electrification, the nature of charge carriers, the spacing between contacting bodies; the effect of friction, pressure, surface roughness and its structure, the role of the external electric field, the effect of humidity; the processes of charge dissipation (emission into the atmosphere, conductivity, electron emission, ion desorption, gas discharges); the ways to suppress the static electricity, and other phenomena.

These problems show the complexity of the mechanism of static electrification, which is the result of close connection between the physical and the dynamic processes typical for such complex compounds as polymers. It is enough to mention that so far no unambiguous determination of the nature of charge carriers in polymers has been reached in a number of cases.

The nature of charged particles crossing the interface between two contacting bodies is essential in determining the mechanism of static electrification. For metal-metal contact the transfer of electrons is most probable as in this case they are free charge carriers. A similar mechanism acts in the contacts between semiconductors and dielectrics, though it is reported that the ion transition is possible at the contact of two dielectrics or a dielectric and a metal [74].

Peterson [79] has reported that the charge carriers between a metal and dielectric can be both ions and electrons, the latter being localized in the dielectric at some energy levels, such as capture levels.

Wagner [80], in his studies of the mechanism of charge transition during contact electrification of SiO_2, has demonstrated that electrons travel from the metal to the quartz surface under the influence of the contact potential difference of about 1 V.

A linear dependence of the amount of charge injected into a polymer at friction on the electron work function of a contacting metal has been reported. It proves the fact of the electron exchange between metals and polymers. Yet, though it is noted that the tribocharge generally tends to increase with the work function for metals, it is also believed that there is no simple universal relationship between the amount of the charge injected into the polymer and the work function of the contacting metal.

Triboelectrification and mass transfer at the sliding of polymers on a paper strip are correlated. It was found that materials with lower wear resistance have a stronger tendency to acquire the negative charge.

Several authors have conjectured that the electrification may depend on the transfer of ions from one surface to another. For example, according to Cornfield [81], solid dielectrics possess their own electrical charge produced by the charged defects in the crystalline lattice. This charge is compensated by selective adsorption of ions of the opposite sign. Friction destroys this compensation causing triboelectrification.

Several attempts were made to explain the mechanism of electrification of dielectrics at dynamic and static contacts using the theory of energy zones and the ideas of volume and surface states [77, 67, 66]. According to these ideas, the presence of the contact lowers the surface energy barriers both for metals and dielectrics. Thermal and mechanical energy cause electrical charges to cross these barriers and localize both on the surface state (SS) or on the volume state (VS). Krupp [67] has specifically shown that the surfaces of dielectrics become charged due to the filling of SS with charge carriers. The amount of polymer charge generated by friction depends on the density and the energy spectrum of the SS. Charge transfer from the surface into the volume states at dynamic contact may be due to the motion of molecules. It has been demonstrated for PS that depending on the work function of a metal, the charge carriers localize either on the surface (in case of metals with high work function) or in the volume (for metals with low work function) states [82]. The depth of the layer with the filled SS reaches 4 to 6 μm.

Thus, the experiments demonstrated that the charge generated by contact electrification can localize both on the volume and surface states depending on the nature of contacting bodies, interaction conditions, and the environment.

When contact between metal and dielectric is just-established or was just-broken, the rate of electrification or the magnitude of appearing charge is a function of the contact potential difference, the nature and concentration of charge carriers, electrophysical characteristics of the dielectric, the contact gap, external conditions, and the contact area and speed of the establishing or breaking of the contact .

Rose and Yard have obtained a relation between the charge magnitude and the size of the "real" contact area. They demonstrated that the direction of charge transfer is defined by the dielectric, not by the type of a metal. They attributed the charging polarity to the dielectric position in the triboelectrical series, thus concluding that the electrification occurs due to the potential difference between two contacting surfaces.

Figure 1.3 shows the diagram of evolution of the volume charge in the contact between the metal (Me) and the n-type semiconductor (SC) (here W_m and W_s are their work functions, and often one has $W_m > W_s$). As soon as the contact is formed, electrons start to "flow" from SC to Me. The region of SC adjacent to the interface becomes electron-depleted, positively-charged, and "attractive" for electrons of the metal. This results in a double layer formation when the equilibrium Fermi level is achieved. Bredov and Kshemyanskaya [84] have investigated the contact electrification process of two bodies and came to the similar conclusion.

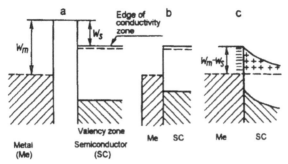

Figure 1.3. Formation of volume charge layer at contact metal (Me) – n-type semiconductor (SC): a, Me and SC before contact; b, Me and SC coming into contact; c, Me – SC contact in equilibrium (Fermi level is shown by dashed line).

When charges localize on the dielectric surface, charge-free or negatively charged zones coexist with the positive surface charge, *ie* the "mosaic" pattern is formed. Similar data are reported for polymer electrets.

The charge polarity depends significantly on the material surface state [84]. In particular, polymer films containing less than 0.1% water acquire a strong negative charge, whereas over 1% water is absorbed, the polarity varies with the positive charge predominance.

Generally, a long-standing interest for the static electrification should be noted. A great number of studies has been devoted to the dependence of the electrification on the contact proximity and its area. The contact proximity defines the maximum distance necessary for charge carriers to reach the opposite surface in the amounts sufficient to be measured before separation. Harper [85] has studied this problem and has shown that the proximity is equal to 25Å for electrons and, apparently, is less for ions.

The effect of the contact proximity on the charge transfer is evidenced by the dependence of charge magnitude on pressure (the contact load), especially at triboelectrification, as reported [74,86,87].

In the macroscopic approach, the notion of contact proximity is inseparable from the definition of the type of the contact, whether it is pointlike or covers some area. Usually, the pointlike contact is an elastic contact between solids. To achieve contact along the part of some plane it is necessary for one of the bodies to be "plastic" at the moment the bodies touch. It is worthwhile to recollect the classical work of Bowden and Tabor [88] in this connection, in that they demonstrated that even a small load would produce high pressure in separate contact points resulting in plastic flow around the points and increase of the real contact area as pressure grows. Strict examination taking into account the "molecular roughness" has shown that any real contact always is pointlike [39].

Friction electrophysics pose many problems in the area of terminology. Peterson and Wagner [79,80] apply the term "contact electrification" instead of "triboelectrification" or "friction electrification." True triboelectrification occurs at "asymmetric" friction of two identical substances, which is a rare case. However, for brevity and convenience of identification and, also, taking into account the complexity of friction process, the general term "triboelectrification" is widely used.

The extent of polymer electrification at friction strongly depends on the sliding velocity and temperature. The sliding velocity effect on the surface charge magnitude has several aspects. Friction of a copper piece against Teflon at speeds up to 20 mm/s has shown no relationship between the Teflon tribocharge and the sliding velocity [89]. The dependence of the surface charge density on the sliding velocity has an extreme pattern produced by the thermal motion of polymer segments. The polyethylene terephthalate (PETP) surface charge at friction against metals may alter its sign with the increase of the sliding velocity.

Sinohara [90] has demonstrated that the polymetylmethacrylate (PMMA) and PS electrostatic potentials at friction interaction with iron powder in the vacuum are independent of temperature. Meanwhile, Sasaki [91] has discovered a strong dependence of the potential on the temperature with the maximum amount of charge in the vicinity of the glass transition temperature within 293–393 K for a number of polar polymers when they rubbed against platinum.

The charge leaking losses are small even at elevated temperatures. The charge density alters insignificantly with temperature for moderate variations of plastic properties of polymers. Therefore, the effect of the high elasticity of the material in the contact zone on the amount of the electrification by friction is explained by the interphase adhesion mechanism. Ohara [92] has studied friction of various couples of polymer films to define the dependence of electrification on temperature. He has noted that the charge density generated on polymer film surfaces at constant sliding velocities reached the maximum in the vicinity of the glass transition temperature of, at least, one polymer in the couple. Moreover, the maximum has shifted into the range of high temperatures and its magnitude reduced as the sliding velocity increased. Appearance of the maximum is attributed to the alteration in the molecular chain mobility as the contact temperature increases.

Studies have shown connection between the chemical structure of a dielectric and its ability to be electrified at tribocontact with other bodies. Lagunova and Vasilenok [93] have revealed that a butyral group, for example polyvinyl butyral (PVB), as a substitutor in a polymer significantly worsens the surface antistatic properties. A relationship has been established between the antistatic effect and the polymer structure [94].

With increasing surface roughness, the tendency toward negative tribocharging is reinforced for practically all the materials [95]. This effect is impossible to explain simply by the fact of alteration of the contact area. Perhaps, friction increases the density of defects (traps of electrons) appearing due to a rough machining.

Analysis of friction processes from the point of view of electron catalysis theory [58,78,83,96] indicates that wear debris can be considered as impurities, or solid catalysts, on the surface.

The following mechanism of tribocharging is probable: appearance of wear debris, adsorption (during contact), surface charging, appearance of the volume charge (field), shift of the Fermi level under the field effect, and modification of the surface properties (adsorbability and catalytic activity, electric conductivity, electron work function). The cycle can repeat itself if impurities (wear debris) have donor or acceptor properties. It is known that the acceptor impurity shifts the Fermi level down, which accelerates the donor reactions governed by holes. The donor impurity shifts the Fermi level up causing acceleration of the reactions governed by electrons, the so-called acceptor reactions. As an example, let's note that oxygen is a strong acceptor, H_2O is a donor, C_3H_7OH is a donor, and C_3H_6 is a donor. Based on the specific catalytic activity with respect to the hydrogen oxidation the metal oxides are arranged in the following manner:

$t = 573K$: $Co_3O_4 > CuO > MnO_2 > NiO > Fe_2O_3 > ZnO > Cr_2O_3 > V_2O_5 > TiO_2$;
$t = 423K$: $Co_3O_4 > CuO > MnO_2 > NiO > Fe_2O_3 > Cr_2O_3 > V_2O_5 > ZnO > TiO_2$.

A great number of molecules and functional groups are known to possess donor or acceptor properties governing the polymer properties: hydroxyl – OH, carboxyl – COOH, nitryl – CN, benzol ring, haloids, chlorine – C, fluorine – F, and others. The polar groups (COOH, OH, NH_2) actively react with the functional groups of adjacent chains. Groups

COOH, OH, and NH_2 strongly increase the adhesion of polymers to various surfaces. A double link (for example, $>C=C<$) plays an important part, especially in branched chains. Hydrogen links are also active [97].

It was mentioned above that triboelectrification of polymers is accompanied, as a rule, by the electret effect. Studies of electrets have a rather long history. As early as the nineteenth century, Heavyside has predicted the existence of dielectrics with constant polarization, analogs of the constant magnets, which he termed electrets. Strictly speaking, the electrets are dielectrics continuously creating an electrostatic field in the surrounding space by pre-electrification or polarization. Eguchi [73,98] has produced the first electret by melting and solidifying carnauba wax in a constant electrical field. The two surfaces acquired electrical charges that gradually changed the polarity. Initially, the charges of the surfaces opposed those of the electrodes, "heterocharge", but later they acquired the same signs as charges on the electrode "homocharge."

According to the modern phenomenological theory of the electret state, an electret is characterized, on the one hand, by the effective surface charge with the density $\sigma_{eff}(t)$ that is usually a difference between a slowly decreasing residual polarization $p(t)$ and a free charge $\sigma_p(t)$:

$$\sigma_{eff}(t) = \sigma_p(t) - p(t) . \tag{1.23}$$

On the other hand, when the electret is thermally depolarized, two shift maxima appear, one with the sign corresponding to the residual polarization destruction and the other corresponding to the free charge relaxation in the volume of dielectric. The "effective" surface charge density σ_{eff}, unlike the "true" density determining the charge density on the electret surface, characterizes both the surface charge and the volume charge distribution. The electret charge variation as a function of time after polarization is determined by the time variations of the residual polarization, free charges migration, and appearance of relaxation polarization . In accordance with these ideas [73] :

$$\sigma_{eff}(t) = \sigma_0'(0)exp\left(-\frac{\tau}{\tau_m'}\right) + \frac{\varepsilon_\infty}{\varepsilon_{st}}\sigma_p(0)exp\left(-\frac{t}{\tau_m}\right) +$$
$$+\left[\frac{\varepsilon_{st}-\varepsilon_\infty}{\varepsilon_{st}}\sigma_p(0) - p_0\right]exp\left(-\frac{t}{\tau}\right). \tag{1.24}$$

The first term in equation (1.24) relates to the capture of injected carriers by tiny traps near the dielectric surface. Its role is especially important for static electrification and charging of corona discharge. The second term relates to the capture of carriers by deep traps with high activation energy. The third item reflects the residual and relaxation polarization of the electret.

The electret state has been investigated for many solid dielectrics: inorganic solids (ion crystals), liquid crystals, biological materials, and polymers. Polymers are the most promising electret materials due to their high dielectric properties, mechanical strength, elasticity, heat resistance, and easy fabrication.

There are different types of electrets, depending on the technology of fabrication [99]: thermoelectrets, cryoelectrets, radioelectrets, corona electrets, electroelectrets, chemoelectrets, mechanoelectrets, magneto-electrets, metal-polymer electrets, and polyelectrets.

Emergence of the electret state in polymers may be caused by the accumulation of free charges resulting from injection processes and bounded charges due to polarization (residual polarization).

Lushchejkin [73] has studied the emergence of the electret state in polymers during tension and compression. He has shown that for the polar polymers polyvinylchloride (PVC) and PMMA containing polar groups in lateral chains, the formation of the electret state is due to orientation of these groups under mechanical effects (mechanoelectrets). Nonpolar polymers PE, PS, and polypropylene (PP) form the electret state due to the injection of charge carriers from metallic rolls into the polymer material ("technological" electrets).

Emergence of electret state in dispersed polymers dynamically contacting with a metal is demonstrated [11]. We have analyzed interrelations between structural transformations of the particles, their morphological modification, the rate of contact electrification, and the emergence of the electret state.

A manifestation of the electret effect during friction of polymer composites with metal fillers has been demonstrated [100]. An abnormal growth of the residual tribocharge in response to the dispersion of the filler and change in its concentration has been found and the relationship between the charge density and the friction coefficient has been shown.

Formation of the electret state alters the structure and properties of polymers. Polyamide (PA) after polarization in the electrical field with the intensity from 50 to 500 kV/cm acquires a greater degree of crystallinity. When polycarbonate is converted into the electret state, the density of specimens somewhat increases. This is explained by the formation of ordered structures of higher density in the electrets. A greater structural ordering and a higher density of the ordered regions have been confirmed by infrared spectroscopy and radiographic analysis for the electrets of polyethylene terephthalate (PETP) and polyacrylonitrile (PAN).

Structural transformations of polymer materials at electretization alter their physical and mechanical properties. More ordered and compact-structure obtained by thermal electretization results in noticeably higher strength of polymers. The destroying elongation stress and the ultimate forced elasticity increase by 15 to 50 percent and the relaxation time of mechanical stresses increases several times in polarized films of PETP, PA, polycarbonate (PC), and polytrifluorine chlorine ethylene (PTFCE) [73]. The strength of polar materials electrets depends strongly on the polarizing field intensity. The nonpolar PE does not exhibit any changes in its strength after polarization. This is confirmed by the influence of the dipole (oriented) polarization on the physical and mechanical properties of polymers.

The electret field alters the processes of absorption and diffusion on polymer surfaces: absorption and diffusion of liquids and gases into metal-polymer electrets reduces due to electrostatic polarizing interaction between the absorbat and absorbent.

Thus, the electrification of polymers at friction results in the inception and development of the electret state. This fact is confirmed by the results of electrophysical studies of contact electretization of dispersed polymers in which a long-lasting electret state has been identified. It is characterized by a significant absolute magnitude of charge (10^{-8} C/cm^2) capable to maintain its stability for a long time (hundreds – thousands of hours), to reverse the charge sign and to retrieve the charge when the air humidity is reduced. It has been shown that the electret state formation is produced by injected charges, polarization and high temperatures in the microcontact. It is affected by the structural transformations in polymeric particles, their morphological changes, and the rate of contact electretization [101,102].

Identification of major interdependent attributes (the charge of 1–10 μC/m^2 in specimens subjected to friction, an existence of the charge for up to 160 days or more and the form of the spectrum of thermally stimulated current (TSC)) has allowed to reveal the electret effect at friction of polymers in liquid media, both dielectrics and electrolytes [103]. It has been mentioned that during friction in liquids with various electrical nductivities, the effect of adsorption on the magnitude of the charge generated by friction has a more general character.

As a consequence, the electret state of polymers produced by friction affects their physical and chemical processes. Particularly, the adsorption evolves mainly on electrically active centers at the surfaces of solids. Therefore an emergence of electret state results in increased activity of the polymer in tribochemical processes [104].

When the electret state is initially induced in a polymer, it becomes possible to control the parameters of friction interaction. The radiographic structural analysis and infra-red (IR) spectroscopy have shown 2 to 5 percent increase in crystallinity and a corresponding reduction of the crystallite size for polyethylene coatings subjected to thermal electretization [105,106]. The thermally stimulated current technique has revealed that, during friction of a composite with the electret filler, the electret field adds to the field produced by triboelectretization, *ie* the electret triboelectrification undergoes a superposition. By altering the direction of the field intensity vector produced by the electret filler particles, one can weaken or strengthen the field appearing at triboelectrification, *ie* to control the friction contact parameters. In particular, the introduction of negatively charged PTFCE filler into the polyethylene matrix reduces the friction force.

In this connection, it is important to treat the electret state of polymers induced by friction as a process of energy accumulation by the electron states ionization. Undoubtedly, the relaxation of the electret charge is typical for electrets and this process can be considered as the release of accumulated energy. However, the electret is known to be a dielectric possessing a "quasiconstant" electric charge.

The idea of energy accumulation during friction in an electret state has received some support. Energy accumulation, and not only dissipation, has not yet received strict mathematic description, but phenomenological arguments support it. The electrophysical theory of friction of dielectrics should be developed using the results achieved in molecular and oscillatory models of friction taking into account the formation and changes of the electret state in dielectrics. The electret effect in polymer dielectrics can be a link combining the major postulates of the molecular and elementary friction theories based on the oscillatory model. These topics are treated in more detail during the analysis of original studies described in the subsequent chapters.

REFERENCES

1. Belyi, V.A. and Sviridenok, A.I. *Soviet Journal of Friction and Wear*, Allerton Press, N.Y., 1987, **8**, no.1, 1–16.
2. Jost, H.P. *Soviet Journal of Friction and Wear*, Allerton Press, N.Y., 1990, **11**, no. 1, 125–133. In 'Tribology – Solving Friction and Wear Problems'. 10th Int. Colloquium, 1996, Esslingen, Germany, p.1.
3. Belyi, V.A. and Myshkin, N.K.. in 'Tribology in the USA and the Former Soviet Union: Studies and Applications', Allerton Press, N.Y., 1994. p.3.
4. Heinicke, G.. 'Tribochemistry', Munich, 1984.
5. Ajnbinder, S.B. and Tyunina, E.L.. 'Introduction into the Theory of Friction in Polymers' (in Russian), Riga, 1978.
6. Vasilenok, Yu.N. 'Protection of Polymers Against Static Electricity' (in Russian), Leningrad, 1981.
7. Vinogradov, G.V. 'Encyclopedia of Polymers' (in Russian), **1**, Moscow, 1972, p. 198.
8. Kestelman, V.N., Evdokimov, Yu.M. and Schindel-Bidinell, E. *Plaste und Kautschuk*, 1992, **11**, p. 375–376.
9. Ponomarenko, A.T., Shevchenko, V.G., Kryzhev, V.G. and Kestelman, V.N. *Int. J. Polymeric Mater.*, 1994, **25**, 207-226.
10. Sviridenok, A.I., Myshkin, N.K., Kalmykova, T.F. and Kholodilov, O.V. 'Acoustic and Electrical Methods in Triboengineering', Allerton Press, New York, 1989.
11. Klimovich, A.F. *Proc. of the BSSR Academy of Science* (in Russian), 1980, **24**, no.3, 238–241; 1986, **30**, no. 12, 1087–1090.
12. Deryagin, B.V., Krotova, N.A. and Smilga, V.P. 'Adhesion of Solids' (in Russian), Moscow, 1973.
13. Evdokimov, V.D. and Semov, Yu.I. 'Exoelectron Emission at Friction' (in Russian), Moscow, 1973.
14. Gorokhovskii, G.A. 'Polymers in the Treatment of Metals'(in Russian), Kiev, 1975.
15. Dmitrieva, T.V., Logvinenko, P.N., et al. *Soviet Journal of Friction and Wear*, Allerton Press, N.Y., 1989, **10**, no. 4, 31–36.
16. Lebedev, L.A. and Georgievskii, G.A. 'Electrochemical Processes in Friction and Their Application for Inhibition of Friction' (in Russian), Odessa, 1973, p. 21.
17. Zaporozhets, V.V. in 'Problems of Friction and Wear' (in Russian), Issue 2, Kiev, 1972, p. 77.
18. Konovalov, E.G., Borisenko, L.B. and E.N. Voznesenskaya *Proc. of the BSSR Academy of Science*, Phys. and Eng. Ser. (in Russian), 1974, **3**, 43–47.
19. Poduraev, V.N., Barzov, A.A., Goldobin, N.D. and Login, V.P. *Machine Building Bulletin*, (in Russian), 1981, **4**, 15–19.
20. Shpenkov, G.V. 'Physicochemistry of Friction' (in Russian), Minsk, 1991.
21. Postnikov, S.N. 'Electrical Phenomena in Friction and Cutting' (in Russian), Gorkiy, 1975.
22. Ter-Oganesyan, V.I. 'The Effect of Triboelectrification on Wear Resistance of Metal-Polymer Couples'. Ph.D Thesis (in Russian), Rostov-on-Don, 1989.
23. Garkunov, D.N. 'Triboengineering' (in Russian), Moscow, 1983.
24. Piloyan, G.O. in 'Proc. of the All-Union Symp. on the Problems of Thermal Analysis Techniques' (in Russian), Moscow, 1968, p.35

25. Belyi, V.A., Sviridenok, A.I., Petrokovets, M.I. and Savkin, V.G.. 'Friction and Wear in Polymer-Based Materials', Pergamon Press, New York, 1982.
26. Bilik Sh.M. 'Metal-Plastic Friction Pairs in Machinery' (in Russian), Moscow, 1965.
27. Sviridenok, A.I. *Tribology International*, 1991, **24**, no. , 37–44.
28. Evdokimov, Yu.A., Sanches, S.S. and Sukhorukov, N.A. *Mechanics of Polymers* (in Russian), 1973, **3**, 520–525.
29. Dubinin, A.D. 'Energy of Friction and Wear of Machine Parts' (in Russian), Moscow, 1963.
30. Shcherbinin, A.I. and Geller, Z.I. *Treatment of Materials with Electrons*, 1967, **16**, no. **4**, 67–69.
31. Eychhorn, R.M. *IEEE Transaction on Electrical Insulation*, 1976, **12**, no.1, 1137–1157.
32. Akhmatov, A.S. 'Molecular Physics of Boundary Friction' (in Russian), Moscow, 1963.
33. Bowden, F. and Tabor, D. 'The Friction and Lubrication of Solids'. Pt.II, Clarendon Press, Oxford, 1964.
34. Kragelskii, I.V. 'Friction and Wear' ,Elmsford, 1982.
35. Deryagin, B.V. 'What is Friction?' (in Russian), Moscow, 1963.
36. Deryagin, B.V. and Krotova, N.A.. 'Adhesion' (in Russian), Moscow, 1949.
37. Helmpolts, H. *Ann. Phys.*, 1979, **7**, 337.
38. Deryagin, B.V., Krotova, N.A. and Smilga, V.P. 'Adhesion of Solids' (in Russian), Moscow, 1973.
39. Sviridenok, A.I., Chizhik, S.A. and Petrokovets, M.I. 'Mechanics of Discrete Contact' (in Russian), Minsk, 1990.
40. Deryagin, B.V., Abrikosova, I.I. and Lifshits, E.M. *Physical Sciences Bul.* (in Russian), 1958, **64**, no. 3, 493–528.
41. Landau, L.D. and Lifshits, E.M. 'Electrodynamics of Continuous Media' (in Russian), Moscow, 1959.
42. Bufeev, V.A. in 'Electrochemical Processes in Friction and Their Application for Inhibition of Friction' (in Russian), Odessa, 1973, 7–10.
43. Anisimova, V.I., Klyuev, V.A., Vladykina, T.N., et al. *Proc. of the USSR Acad. of Sci.* (in Russian), 1977, **233**, no. 1, 140–143.
44. Anisimova, V.N., Deryagin, B.V. and Toporov, Yu.P. *Colloid Journal* (in Russian), 1984, **XVI**, Iss. 5, 1039–1041.
45. Deryagin, B.V., Krotova, N.A. and Knyazeva, N.P. *Proc. of the USSR Acad. of Sci.* (in Russian), 1974, **215**, 1078–1081.
46. Krotova, N.A. and Khrustalev, Yu.A. *Colloid Journal* (in Russian), 1974, **36**, 480–483.
47. Lebedev, P.I. 'Selected Papers' (in Russian), Moscow, 1949.
48. Lifshits, E.M. *Proc. of the USSR Acad. of Sci.* (in Russian), 1955, **100**, no. 5, 879–881; 1954, **97**, no. 4, 643–645.
49. Belyi, V.A., Sviridenok, A.I. and Smurugov, V.A. *Surface. Physics, Chemistry, Mechanics*(in Russian), 1986, **3**, 130–140.
50. Smurugov, V.A., Senatrev, A.I., Savkin, V.G., Biran, V.V. and Sviridenok, A.I. *Wear*, 1992, **158**, 61–69.
51. Rogachev, A.V., Buj, M.V., Pleskachevkii, Yu.M. and Lavrentjev, V.V. *Soviet Journal of Friction and Wear*, Allerton Press, N.Y., 1989, **10**, no. 6, 52–56.

52. Rogachev, A.V., Buj, M.V. and Pleskachevskii, Yu.M. *Soviet Journal of Friction and Wear*, Allerton Press, N.Y., 1989, **10**, no. 4, 112–116.
53. Bartenev' G.M. 'Strength of Polymers and their Failure' (in Russian), Moscow, 1984.
54. Tamm, *IE. Sov. Phys.* (in Russian), 1992, **1**, 733–737.
55. Shockley, W. *Phys.Rev.*, 1939, **56**, 317–323.
56. Bardeen, J. *Ibidem*, 1947, **47**,. 717–727.
57. Shockley, W. and Pearson, G.L.. *Ibidem*, 1948, **74**, 232–233.
58. Kiselev, V.F. and Krylov, A.V. 'Electron Phenomena in Adsorption and Catalysis on Semiconductors and Dielectrics' (in Russian), Moscow, 1979.
59. Parkinson, G.M., Thomas, J.M. and Williams, J.O. *J.Phys.*, 1974, **7**, 310–313.
60. Aris, F.C., Brodribb, J.D., Hughes, D.M. and Lewis, T.J. *Sci. Papers, Inst. Org. Phys. Chem.* Wroclaw Tech. Univ. 1974, **7** , 182–187.
61. Vannikov, A.V., Matveev, V.K., Sichkarov, V.P. and Tyutnev, A.P. 'Radiation Effects in Polymers. Electrical Behavior' (in Russian), Moscow, 1982.
62. Baru, V.G. and Valkenshtejn, F.F. 'Effect of Surface Irradiation on Behavior of Semiconductors' (in Russian), Moscow, 1978.
63. Distler, G.I., Vlasov, V.P. and Gerasimov, Yu.M. 'Decoring Surfaces of Solids' (in Russian), Moscow, 1976.
64. Kao, K. and Huang, V. 'Transfer of Electrons in Solids', Vol.2 (Russian translation), Moscow, 1984.
65. Rzhanov, A.V. 'Electron Processes on Semiconductor Surfaces' (in Russian), Moscow, 1971.
66. Klimovich, A.F. and Mironov V.S. *Soviet Journal of Friction and Wear*, Allerton Press, N.Y., 1985, **6**, no.5, 18–26; no.6, 52–58.
67. Krupp, H. *Static Electrification Conf. Ser.*, no. 11, Inst. Physics, London, 1971, 1–15.
68. Fabish, T.J., Saltsburg, H.M. and Hair, M.L. *Sci. of Appl. Phys.*, 1976, **47**, no.3, 930–939.
69. Cessler, G. in 'Electrets', (Russian translation), Moscow, 1983, p. 8–104.
70. Bauser, H. *Kunststoffe*, 1972, **62**, 192.
71. Seggern, H. *Journal Appl. Phys.* , 1979, **50**, 2817.
72. Butyagin, P.Yu. *Proc. V Symp. on Mech. Emission and Mech. Chem. of Solids* (in Russian), Tallinn, 1975, 10–11 and 70.
73. Lushchejkin, A.G. 'Polymeric Electrets', (in Russian), Moscow, 1976.
74. Loeb, L. 'Static Electrification' , Springer Verlag, 1958.
75. Lennard, P. *Ann. Phys.*, 1892, **46**, 584.
76. Coehn, A. *Wiedeman Annalen.* , 1898, **64**, 217.
77. Gvezekhus, N.A. *Journal of Rus. Phys. Chem. Soc.* , 1902, **34**, 367–371.
78. Richards, N. *Physical Rev.* , 1910, **22**, 2.
79. Peterson, J.W. *Appl. Phys.* , 1954, **25**, no. 40, 501–504.
80. Wagner, P.E. *J. Appl. Phys.*, 1956, **27**, 1301.
81. Cornfield, M.N. *Theor.Phys.* , 1975, **17**, no. 8, 2516–2517.
82. Fabish, T.J., Saltsburg, H.M. and Hair, M.L. *Sci. of Appl. Phys.*, 1976, **47**, no.3, 930–939.
83. Bredov, M.M. and Kshemyanskaya, I.Z. *J. Theor. Phys.* (in Russian), 1957, **27**, 923–929.

84. Schumann, W. *Plaste und Kautschuk* 1963,**10**, no.9, 526–531.
85. Harper, W.R. 'Contact and Frictional Electrification', Oxford, 1967.
86. Staroba, N. and Shimordi, I. 'Static Electricity in Industries' (in Russian), Moscow, 1960.
87. Lewy, W.W. *SPE Journal*, 1962, **18**, no. 10, 1288–1290.
88. Bowden, F.P. and Tabor, D. 'Friction and Lubrication of Solids' (Russian translation), Moscow, 1968.
89. Wahlin, A. and Beckstron, G.. *Appl. Phys.*, 1974, **45**, n 5, 2058–2084.
90. Sinohara, I., Tsupki, H. and Tsutida, E. Koche *Kachaku Dzassu* (Russian translation), 1970, **33**, no. 7, 1460–1467.
91. Sasaki, K. and Fuiko, S. *Kobunsi Kachaku* (Russian translation), 1971, **28**, no. 313, 415–422.
92. Ohara, K.. *J. Electrostatics*, 1978, **4**, 233-246.
93. Lagunova, V.I. and Vasilenok, Yu.I. in 'Static Electricity in Polymers' (in Russian), Leningrad, 1968, p. 95.
94. Konoplev B.A. and Vasilenok, Yu.I. in 'Static Electricity in Polymers' (in Russian), Leningrad, 1968, p. 86.
95. Vladykina, T.M., Deryagin, B.V. and Toporov, Yu.P. *Surface* (in Russian), 1984, **9**, 149–151.
96. Enikeev, E.Kh. in 'Physics and Physico-chemical Catalyses' (in Russian), Moscow, 1960, p. 85.
97. Krotova, N.A. and Morshchova, L.P. in 'Investigation in the Domain of Surface Forces' (in Russian), Moscow, 1961, p. 83.
98. Gubkin, A.H. 'Electrets' (in Russian), Moscow, 1978.
99. Lushchejkin, G.A. 'Methods of Investigation of Electrical Behavior of Polymers' (in Russian), Moscow, 1988.
100. Mironov, V.S. and Klimovich, A.F. *Proc. of the BSSR Acad. of Sci.* (in Russian), 1986, **30**, no. 8, 724–727.
101. Klimovich, A.F. and Mironov, V.S. *Soviet Journal of Friction and Wear*, Allerton Press, N.Y., 1981, **2**, no. 3, 128–131.
102. Belyi, V.A., Pinchjuk, L.S., Klimovich, A.F. and Guzenkov, S.I.. *Proc. 5th Int. Cong.on Tribology*, Helsinki, 1989, 276–281.
103. Guzenkov, S.N. *Soviet Journal of Friction and Wear*, Allerton Press, N.Y., 1990, **11**, no. 1, 151–154.
104. Belyi, V.A., Klimovich, A.F. and Mironov, V.S. *Proc. of the BSSR Acad. of Sci.* (in Russian), 1982, **26**, no. 1, 39–42.
105. Goldade, V.A. and Pinchuk, L.S. 'Electret Polymers: the Physics and the Material Science' (in Russian), Minsk, 1987.
106. Klimovich, A.F. and Guzenkov, S.I. *Soviet Journal of Friction and Wear*, Allerton Press, N.Y., 1989, **10**, no. 5, 6–11.

CHAPTER 2

EXPERIMENTAL TECHNIQUES OF INVESTIGATING ELECTRICAL PHENOMENA IN THE FRICTION OF POLYMERS

Electrical phenomena accompany and initiate the majority of chemophysical processes occurring in the friction of polymers. Electrophysical techniques of investigating the processes of frictional interactions are the most advanced. In addition to the traditional current, probing, and emission techniques, the study of the friction of polymers requires new techniques, in particular those based on the decoration of surfaces and the electret-thermal analysis.

2.1. CURRENT MEASURING AND PROBING TECHNIQUES OF INVESTIGATING TRIBOELECTRIFICATION

The electrophysical processes in the friction of polymers require measurements of electrical parameters, such as electrification current, the potential of a metal-polymer friction couple, the electron work function, the contact potential difference, exoemission of electrons, and so forth. Measurements of the potential or the electrification current are most frequently used. The reason the potential serves as a parameter characterizing the polymer electrification in friction is that each metal-polymer couple under certain conditions has a definite total charge reflecting the tendency of a couple to generate and accumulate electrical charges in friction [1]:

$$\sigma = \frac{Q}{S}.$$
(2.1)

Here S is the contact area, Q is the magnitude of the generated charge, which can be determined from the following expression

$$Q = CU, \qquad (2.2)$$

where C is the capacity of a system; and U is the potential.

When C and S are constant, the values Q, U can serve as the electrophysical characteristics of a dielectric in accordance with the law of saturation.

Bilik and Tsurkan [1,2] analyzed several experimental designs for measuring the potential (Fig. 2.1). These designs allow one to investigate the triboelectrification parameters of metal-polymer couples in friction as well as the effect of a measuring device upon the extent of electrification of polymers. They demonstrated that the total potential for the majority of metal-polymer couples can be determined from the expression

$$U = Kp\tau \tan\alpha \qquad (2.3)$$

where K is the factor describing the polymer material; p is the load; τ is the time of appearance of the potential; and $\tan\alpha$ is the rate of its appearance.

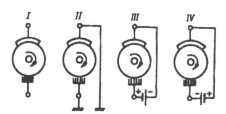

Figure 2.1. Electrical circuits of metal-polymer friction couples [1]: I, insulated rubbing bodies; II, "grounded" bodies; III, external voltage source connected in series; IV, counter-connected source

The approach of Rogers [3] and Shashoua [4] employs indirect measurement of the electrification potential as the basic parameter characterizing the electrification behavior of polymers. First, the charging is induced by the crown charge, then, the maximum potential or the rate of charge relaxation, or both at the same time are monitored. These techniques are complicated and lack accuracy because they do not use the direct measurements of triboelectrification of polymers. The design of Kelemen and Pal [34] is appealing because the electrification is produced directly during the friction of polymer specimens. Its essential drawback is the need to connect the rubbing member to the measuring voltmeter each time the measurement is made, so that the capacitance of the system will not distorts the results.

The technique of investigating the triboelectrification of polymers described in [5] employs a device that includes three major units: the electrifying unit with a special motor-generator, a washer and a rubbing member, as well as the measuring unit and the unit for maintaining required temperature and humidity.

The friction rate and the load are monitored and adjusted. The device is placed inside a chamber to isolate it from the environment and to maintain the temperature with the accuracy of ±1 K and relative humidity of the working medium between 10 to 90%±1%. This provides the necessary conditions for good reproducibility of the results.

The following equation of Shashoua [4] was used to describe the discharge process (the charge relaxation)

$$\ln U = \ln U_{max} - (K_1 + K_2)t. \qquad (2.4)$$

Here U is the potential of a specimen after time t; U_{max} is the initial potential produced by the crown charge, (at $t = 0$, ie at the moment the source of the crown charge is activated); K_1 is the charge dissipation constant determined by the specimen's conductivity; and K_2 is the charge dissipation constant in the air.

The set-up in [5] also allows determinination of the charge induction constant K_{chr} . The charging process in this case is described by the equation

$$\ln U = (K_{chr} - K_1 - K_2)t, \qquad (2.5)$$

where K_{chr} is the charging constant similar to the dissipation constants K_1 and K_2 of Shashoua .

This technique of investigating the tribocharging of polymers allows one to estimate the triboelectrical effects and to select effective antistatics, kinetic charging constants and charge dissipation.

It is essential to simultaneously measure the parameters characterizing triboelectrification in friction in order to explain the mechanism of triboelectrification and its effect on friction. Figure 2.2 shows a block-diagram of a suitable set-up. A polymer film contacts another polymer film, *ie* a polymer-polymer couple is formed.

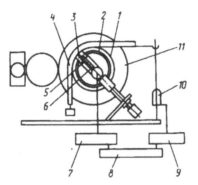

Figure 2.2. Block-diagram of an experimental set-up for measuring triboelectrification and friction parameters of polymer films: 1, polymer film – base; 2, "grounded" copper cylinder; 3, aperture in base and cylinder; 4, tested specimen – polymer film; 5, electrode; 6, electrometric measuring probe; 7, electrometer; 8, recorder; 9, strain gage amplifier; 10, loading device; 11, chamber with controllable temperature and humidity.

The underlying polymer film is wound around the "grounded" copper cylinder, and the hole is pierced in it to fit the hole in the cylinder.

When the cylinder rotates, an electrical charge is generated on the polymer film due to friction of the polymer – base – polymer film. It is registered using an electrical measuring probe that consists of an electrode, a protective ring, and an electrometer.

The friction coefficient is determined from the formula

$$\mu = \frac{1}{Q} \ln \frac{T}{T_0},\qquad(2.6)$$

where T_0 is the initial tension of the polymer film; T is the final tension of the polymer film; and Q is the angle of enclosure of the cylinder by the polymer film.

The tension T was measured using a loading member and a strain amplifier. Signals of the amplifier and the electrometer were recorded. The device was put into a chamber with controllable temperature and humidity.

This set-up served to investigate the effect of temperature and relative sliding velocity on triboelectrification as a response to the frictional behavior of polymers. A significant drawback of the technique and the device is a rather intricate system of calibration and measurements of the electrical charge density, in that the electrostatic potential is registered through the hole in the underlying polymer and the "grounded" metallic cylinder.

The electrophysical processes during the friction of polymers can also be investigated by using current measuring techniques or by registering the triboelectrification current. It is also possible to design a friction stand for synchronous measurements of the friction moment and the current characteristics of the process (Fig. 2.3) [7].

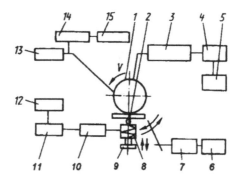

Figure 2.3. Diagram of a stand with the synchronous registration of the moment of friction and the process current characteristics: 1, shaft (counterbody); 2, specimen (partial insert or plate); 3, current meter; 4, amplifier; 5 and 6, recorder; 7, induction pick-up; 8, support; 9, loading device; 10, pick-up; 11, indicator; 12, scale; 13, controller of revolutions; 14, revolution meter; 15, scale.

The shaft acts as a counterbody in the "direct" friction couples or the specimen in the "reversed" couples. It rotates with the speed v, which is controlled within the specified limits for a a fixed number of revolutions. The specimen is made as a partial insert or a plate and is connected to the loading device through the support allowing it to turn in the vertical plane. The moment of friction is registered by the recorder and the induction pick-up. The load is registered by the indicator through the pick-up and the recalculation scale. The friction unit is protected by a "grounded" metal screen.

The friction stand allows investigation of both "direct" and "reversed" friction couples. Polymers can be shaped as block specimens, free films, or coatings on a solid base. It is also possible to measure the temperature in the friction zone using a thermocouple connected to the stationary member of the friction couple.

2.2. THERMAL ANALYSIS OF ELECTRETS.

The techniques employed for investigating the triboelectrification and the appearance of the electret states in polymers during friction can be divided into three groups: the techniques for determining of the surface charge density, the techniques for determining the surface potential, and techniques for measuring the thermostimulated current (TSC).

The effective surface charge density (ESCD) governing the magnitude and the distribution of surface and space charges is one of the major parameters characterizing the degree of electrification of polymers at friction and their electret state.

The effective surface charge density σ_{eff} is equivalent to the potential difference V that is used to describe the behavior of electrets [8]:

$$V = \int_0^L E(x)dx = E_i L = \frac{4\pi\sigma_{eff}L}{\varepsilon} = A\sigma_{eff}. \qquad (2.7)$$

Here L is the electret thickness; E_i is its average internal field; and $A = 4\pi L/\varepsilon$ is the constant multiplier.

The ESCD value and sign are determined by also called contactless compensation technique using a vibrating electrode [8,9]. The technique implies that an external electrical field (V_c) compensates the field in the void between the lower stationary electrode with a flat polymer electret on it and the upper electrode that is vibrated by a sound generator. A direct current source serves as a source of the compensating voltage. When voltage is delivered to the lower electrode the electret field gets compensated.

The ESCD is described by the following formula:

$$\sigma_{eff} = \frac{\varepsilon_0 \varepsilon V_c}{L}[Cl/m^2], \qquad (2.8)$$

where ε is the relative dielectric permeability of the specimen (the electret); $\varepsilon_0 = 8.85 \ 10^{12}$ F/m is the electrical constant; L is the polymer electret thickness, m; V_c is the compensating voltage, V.

The sign of charges on the electret surface can be determined from the sign of the compensating voltage.

One of the most popular techniques of investigating the distribution of charges in the electret body is the so called technique of "consecutive sections" of layers of a material [9,10]. This technique has revealed that the "homocharge" concentrates in a thin layer near the electrode, whereas the "heterocharge" concentrates inside the body of the electret.

The depth of penetration of an injected space charge can be evaluated by comparing the value of the surface potential V_e measured with the help of the compensation technique and those obtained by the TSC integration [9].

When charges are located close to the surface, then

$$V_e = \sigma L / \varepsilon \varepsilon_0,$$

$$\sigma = \int_0^L q(x)dx, \quad \sigma = \int_0^{t_e} i_c \, dx. \tag{2.9}$$

Here i_c is the current during the charging time t_c; L is the electret thickness; x is the depth of charge penetration; and $q(x)$ is the charge density in the layer with thickness dx.

If charges are not near the surface, then, according to the formula (2.9)

$$V_e = \sigma \bar{x} / \varepsilon \varepsilon_0. \tag{2.10}$$

The average distribution of charges in a volume $\bar{x}$ is determined by the expression

$$\bar{x} = \int_0^L xq(x)dx / \int_0^L q(x)dx. \tag{2.11}$$

The distribution of charges in the volume can be determined using the heat pulsing technique [9,11,12]. The experimental set-up used to implement this technique is shown in Figure 2.4.

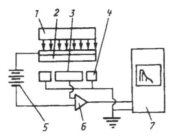

Figure 2.4. Diagram of the experimental set-up for determining the injection depth in electret using heat pulses: 1, pulsing illuminator; 2, electret sample under study; 3 measuring electrode; 4, ring-shaped protective electrode; 5, compensating voltage source; 6, amplifier; 7, recorder.

A pulsing source illuminates the metallized side of the electret during time Δt. The electret is placed into a measuring cell that contains the circular protective and measuring electrode sending signals to the recorder through an amplifier. The electrical field from the "working" insulated side of the electret is compensated from the voltage source.

Extrapolation of V_t to $t \gg \tau$ from $t = 0$ yields $V(t \gg \tau)$, so that from the relation

$$V(t = 0)/V(t \gg \tau) = L/\bar{x} \qquad (2.12)$$

the average disposition of the injected space charge in the electret $\bar{x}$ can be determined.

The center of distribution of charges can be determined in a polymer electret using a combined induction-depolarization technique [10] consisting of two stages. First, the effective surface charge density σ_{eff} on the non-metallized electret surface is measured. Then, any known technique is employed to depolarize the specimen and its full free charge is determined as

$$q_\infty = \int_0^\infty i dt. \qquad (2.13)$$

After that, the center of distribution is determined using the expression

$$\bar{x}/L = \frac{1}{1+[\sigma_{eff}/q_\infty]}. \qquad (2.14)$$

This technique is applicable provided there is no dipole or any other type of space polarization.

The block-diagram of the experimental set-up for the electret thermal analysis of TSC measurements in Figure 2.5 shows a chamber containing an automatic temperature controller for setting and maintaining required heating rate (1.25; 2.5 or 5.0 K/min), and a heated measuring cell.

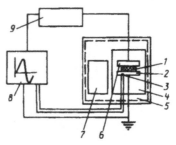

Figure 2.5. Block-diagram of set-up for thermal electret analysis (thermosimulated current measurements): 1, specimen; 2, lower electrode with built-in thermocouple; 3, thermocouple; 4, measuring cell; 5, programmable heater; 6, upper measuring electrode; 7, automatic temperature controller; 8, coordinate recorder; 9, current measuring unit (electromeasuring amplifier).

The specimen is placed between the polished upper and lower measuring electrodes. A thermocouple fixed close to the electret in the lower "grounded" electrode serves for measuring the temperature in the cell. The measuring electrode is connected to the electrometric amplifier, which is highly sensitive to current (10^{-7} to 10^{-15} A). Current variations in response to temperature change are registered with the help of a two-coordinate plotter. The blocking metal electrodes are employed. The measuring electrode and the specimen are separated by the 100 μm thick polytetrafluorethylene (PTFE) spacer.

In case of repeated experiments, the signals from the amplifier and the thermocouple can be directed into the computer through the interface system for saving and processing of the mathematical data: such as computations of the values of parameters and the number of traps, the values of the parameters of relaxation processes, and so forth.

There are several techniques for evaluating the results of TSC measurements, such as the one used used for "close" contacts between the electrodes and the electret [10].

In the general case, the thermostimulated current is given by

$$i = d\sigma / dt = -\sigma(t) / \tau. \tag{2.15}$$

Here τ is the charge relaxation time.

It is assumed that the time τ depends on the temperature ($\tau = A$ exp(-W/RT))and the heating rate b

$$i = \frac{\sigma_0}{\tau} \exp\left(-\int_0^t \frac{dt}{\tau}\right) = -\frac{\sigma_0}{\tau} \cdot \exp\left(-\frac{1}{bA}\int_{T_0}^{T} e^{-W/RT} dT\right). \tag{2.16}$$

The plot of current vs. temperature (or time) has one or several maxima. For $t \ll \tau$

$$|i| = \text{const } e^{-W/RT}. \tag{2.17}$$

From (2.17) follows the formula for determining the relaxation activating energy based on the initial slope at temperatures T_1 and T_2

$$W_i = \frac{19{,}14(\lg i_2 - \lg i_1)}{(1/T_1) - (1/T_2)}. \tag{2.18}$$

The charge of the electret at the moment of time t is

$$\sigma(t) = \int_t^{\infty} i(t) dt,$$

and

$$\tau(t) = \sigma(t) / i(t). \tag{2.19}$$

The values $\sigma(t)$ can be determined by plotting the values of the area under the TSC curve.

After plotting the relation between $\lg \tau(t)$ and $1/T$, the activating energy is derived as

$$W_\tau = \frac{19.14(\lg \tau_2 - \lg \tau_1)}{(1/T_1) - (1/T_2)}. \tag{2.20}$$

It can also be derived using the TSC maximum or T_{max} variations in response to the heating rate changes.

Grossweiner has used the TSC technique to determine the trapping cross section (σ_n) and the frequency of "efforts to escape" (v) using the formula:

$$\sigma_n = \frac{v}{2,9 \cdot 10^{24} \cdot T_m^2}, \tag{2.21}$$

$$\upsilon = \frac{3T'\beta}{2T_m(T_m - T_1)} \exp\left(\frac{U}{kT_m}\right), \tag{2.22}$$

where T_m is the temperature of the TSC maximum; T' is the temperature of the half maximum of the TSC peak measured from the low temperature side.

In the case when the electret contains injected charge carriers of the same sign (specifically, during triboelectrification), the monomolecular recombination dominating in polymers produces the following relation for $i = f(t)$ at the heating rate β [30,35]

$$i = \frac{(n_{i_0}g\delta)^2 \mu\tau}{2\epsilon\tau_0 L} \exp\left(-\frac{U}{kT} - \frac{2}{\beta t_0} \int_{T_0}^{T} \exp\left(\frac{U}{kT}\right) dT\right), \tag{2.23}$$

where τ_0 is the life time of the free charge at $1/T \to 0$; n_{i_0} is the initial concentration of charge carriers; τ is the free charge lifetime; $1/\tau_0$ is the frequency of escapes from traps; δ is the depth of penetration of the carriers into the electret; L is the electret thickness; g is the charge carrier magnitude; and μ is the mobility of charge carriers. The parameters of the injected charge (the concentration of charge carriers), the product of the mobility by the life time of the free charge, and the charge average penetration are determined using the formula

$$n_{i_0} = \frac{\sigma}{q\delta S} \cdot \frac{2L}{\delta}, \tag{2.24}$$

$$\mu\tau = \epsilon \frac{\delta}{L} \cdot \frac{\delta S}{\sigma}, \tag{2.25}$$

$$\delta = \frac{2L(\sigma/S)}{\sigma_{ef} + \sigma/S}, \tag{2.26}$$

where σ is the total   released charge determined by integration

$\sigma = \int_0^{\infty} i(t)dt$; S is the electrode area; and σ_{eff} is the effective surface charge density prior to TSC.

The technique of thermostimulated currents or the technique of thermostimulated charges are the most effective tools for investigating the electret state of polymers and composites. The second technique serves to measure the charge relaxation in the electret state. Because the charge relaxation at room temperature is a long-lasting process, the thermal stimulation is applied by placing the specimen between two electrodes and heating it at a rate 1 to 5 K/min, after which the passing current is recorded. The plot of the current as a function of temperature allows identification of a definite set of peaks on the TSC spectrum and this enables the evaluation of the parameters and the determination of the mechanism of appearance of the electret state. The TSC allows determinination of the major parameters of the relaxation process: the mechanism of relaxation, the time and the energy of the relaxation activating process, the capture parameters, and so forth. The technique possesses high sensitivity and resolution; it enables observation of the relaxation processes in polymers in different phase states and registration of the transition moments. Also, the TSC technique measures a number of factors governing the conditions of fabrication and performance of polymer and polymer-based composites, such as mechanical stresses and deformation, bulk heterogeneity, surface states along the boundaries separating heterogeneous structures, the quality of the metal-dielectric contact, the effect of different types of irradiation, and the processes of adsorption and desorption of gases [13]. The advantage of this technique is the possibility to investigate simultaneously the surface and the volume transformations in polymers under the influence of the external effects. It is especially important for the studies and prediction of performance of polymers and polymer contacts during friction and wear.

The TSC theory is treated in more detail in [8,9,14,15].

2.3. ANALYSIS OF EMISSION PROCESSES

Emission of electrons often serves as a source of electrical signals for various measuring techniques and in various devices for investigating friction between solids [16,17,18]. In particular, the emissions of slow electrons, exoelectron emission (EEE) and mechanoemission of fast electrons, high energy electrons (HEE) are used most often.

The EEE is a stimulated process causing the emission of electrons with the energies below 1 eV. The EEE appears most frequently during machining, disintegration and deformation of materials, friction, irradiation, and so forth. The stimulation mechanism can cause EEE to be optically stimulated (photoemission), thermostimulated (thermoemission), adsorption stimulated (chemoemission), as well as triboinduced or induced by electrical fields.

Kramer [19] has initiated of EEE systematic studies by investigating its temperature dependence. He put forward a hypothesis stating that the EEE is excited by the latent heat of phase transformations. Later, a hypothesis was proposed stating that the EEE mechanism is initiated by chemosorption. Yet, based on its "recombining nature", the EEE was determined to be the emission caused by the restructuring of surface layers and the appearance of an "active" surface [20,21].

In 1953, Deryagin et al. have discovered the emission of HEE [22] by studying the adhesion processes and the appearance of new surfaces due to disintegration of some crystalline solids in the vacuum. Later the HEE emission was found to exist during the disintegration of adhesion and cohesion bonds in the process of friction of dielectrics in the vacuum. The HEE energy reaches 10^4 to 10^5 eV, *ie* it is several orders of magnitude greater than the energy of EEE.

A relation has been reported [23] between the parameters of the flux of emitted electrons and the charge density on the juvenile surface. It has been demonstrated that the higher energies of emitted electrons are due to the electrical fields on juvenile surfaces. The energy of HEE can be determined from the formula

$$E = 2\pi\sigma er, \qquad (2.27)$$

where σ is the charge density; e is the charge of an electron; and r is the charged area radius.

The analysis of expression (2.27) indicates that the HEE emission is possible only from the surfaces on which charged sites have a definite size (the radius r), so that the active surface emits only from individual spots or "active centers."

The autoelectron mechanism (tunnelling of electrons under the effect of the local high voltage fields) is especially attractive hypothesis for explaining the HEE emission [24].

The techniques for experimental studies of the emission phenomena, especially the emission during friction, are treated most comprehensively in [16,25]. Let us consider, for example, the study of the friction in the air using the experimental set-up proposed by Amsler. A ring-shaped specimen of 60 mm in diameter and 15 mm wide is fastened horizontally in the spindle of the experimental apparatus. A detector or an open-ended electron counter protected against vibration, interference, noise and penetration of debris into the operating space, is mounted over the ring. The equipment allows for continuous observation over and registration of the electron emission induced by friction in the air.

An extensive body of the experimental information about characteristics of the emission processes is being accumulated. It shows the broad, practical applicability of the emission processes for studies of various phenomena serving as indicators of the disintegration processes in dielectric materials [16,18,25–29].

2.4. TRIBOELECTRICAL MEASUREMENTS IN VARIOUS MEDIA

There is a growing need to evaluate the effects of the environment in the contact zone on the electrophysical processes in friction, specifically those appearing during friction of polymers in the vacuum.

Figure 2.6 shows the block diagram of the device designed as a "shaft-partial insert" for investigating triboelectrification of polymers under such conditions. The metallic insert with a polymer coating (in the case of the direct friction couple) or a metallic strip (in the case of the reversed couple) are pressed by a loading unit against the roller surface.

The force of friction is registered by strain gauges attached to the resilient beam and the signals are recorded. The system allows creations of moderate or high vacuum and employs various gases. The gas composition in the contact and the qualitative analysis of volatiles released during friction of polymer materials are determined by mass-spectrometry. The mass-spectrometric sensor is arranged directly in the vicinity of the friction zone. The mass spectrometric analysis of released gaseous products is performed using the commercial mass-spectrometer MX-7301Y42 made in USSR.

It is convenient to characterize the degree of electrification of polymers at friction by the current "flow" from one member of the friction pair -

such as the polymer roller, in the case of the reversed pair, or the insulated insert, in the case of the direct pair - through a measuring device to the "ground".

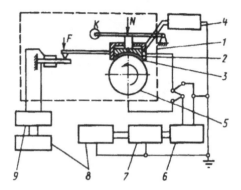

Figure 2.6. Block-diagram of the commercial mass-spectrometer MX-7301Y42 made in USSR for investigating electrification of polymers at friction in vacuum: 1, vacuum chamber; 2, insert; 3, polymer coating; 4, recording potentiometer; 5, roller; 6, current microvoltampmeter; 7, compensation amplifier; 8, recorder; 9, strain-gauge amplifier; K, load; F, friction force; N, sample's load.

The current is registered by an auto-compensated volt-nanoampmeter. The signals are then passed through a compensating amplifier and recorded. Commutation lines with a protective shield are used for current measurements. The friction unit is protected by the "grounded" electrostatic screen. The temperature of the metallic insert is measured by thermocouple, and the results are registered by a recording potentiometer.

A special stand was designed to investigate the appearance of the electret states in polymers during friction against metals in the vacuum (Fig. 2.7).

The friction unit is placed in the working space of the multipurpose vacuum unit. The specimen is secured on a heating table mounted on a resilient beam. A reciprocating metallic indentor is forced into the polymer coating and the vacuum up to 1.3×10^{-2} to 1.3×10^{-3} Pa is created. The force of friction is read by strain gauges glued to the strain beam. A special manipulator creates friction, and after it is released, the indentor is lifted and replaced with the unit for electrophysical measurements conducted at 1mm spacing. Initially, the ESCD is measured using the vibrating electrode by the noncontact compensation technique described above. Switches K_1 and K_2 are set into position 1.

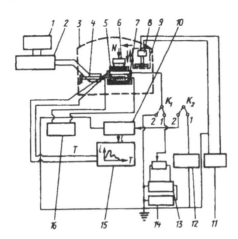

Figure 2.7. Block-diagram of the stand for investigating the appearance of electret state in polymers during friction in vacuum: 1, recorder; 2, strain amplifier; 3, vacuum chamber; 4, resilient beam; 5, specimen; 6, indentor; 7, heating table; 8, measuring unit; 9, vibrating electrode; 10, amplifier; 11, signal generator; 12, electrometer; 13, vacuum unit; 14, electrometer; 15, two-coordinate plotter; 16, programmable temperature controller.

Then, the electret thermal analysis is performed by setting the switches into position 2. A programmable unit produces heating with the linear rate of 3 K/min, while the electromeasuring amplifier serves for recording TSC by sending signals from the thermocouple to the second channel of the two-coordinate plotter.

During friction of metals in liquids, especially in electrolytes, more attention is paid to the electrochemical processes produced by moving the deformable discrete contacts between individual microdefects.

For such studies in liquids, a potentiostatic technique has been developed [18,3] that employs automatic maintenance of a preset electrode potential φ, or its variations, based on a definite program (potentiostatic friction studies allow the investigation of the effect of φ on mechanical friction characteristics and gather a substantial amount of information on the kinetics of electrochemical processes at friction.) Additional techniques include determination of the corrosion rate at friction; linear extrapolation of the real cathode and anode polarization curves measured directly on the specimen, and others.

A potentiometric technique of investigating the electrochemical processes at friction of metals in the electrolyte [31], employs "butt-to-butt" friction by pressing a stationary cylindrical specimen against the experimental specimen. The system is polarized and the three-electrode electrochemical cell performs measurements. The system includes the studied electrode (SE), for which q is measured as compared to the reference electrode (RE) and the auxiliary platinum electrode (AE). These electrodes together form a polarization circuit. The values of φ are measured, and potentiostat and potentiometer record the polarization curves. The system is equipped with a dynamomemeter for measuring the moment of friction. The best valid results can be obtained when the RE is brought directly into contact with the specimen through a hole in a plastic tube.

Figure 2.8 shows the device [32] for investigating the effect of the electrochemical processes on friction and wear of metal-polymer couples in corrosive media. The device is comprised of a tribometer with pendulum, a support and a prism carrying it. The support is a vessel that holds the electrolyte into which the prism rubbing surface is immersed. The pendulum is put into motion with some value of the initial amplitude, whose attenuation is recorded. When pendulum oscillates, a U-shaped core is alternatively introduced into one or the other coil of induction, thus generating the sinusoid signals synchronous with oscillations. The signal is rectified, amplified, and recorded by potentiometer.

A three-electrode design for polarization with a dielectric support employs a metallic bath as a working electrode filled with electrolyte. Platinum wire serves as the auxiliary electrode. The reference chlorine-silver electrode is placed in a vessel connected to the cell via electrolytic key. The electrodes are connected to the terminals of the potentiostat. The metallic bath is polarized in the potentiostatic mode, after which the load is varied, and polarization current, amplitude and duration of oscillations of the pendulum are registered. This device allows investigation of the effect of electrode processes upon the kinetics of wear of metal-polymer couples.

The experimental techniques described offer a wide range of opportunities for evaluating the triboelectrophysical behavior of materials in various media.

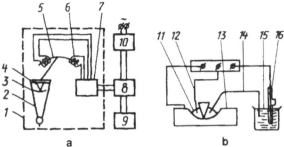

a   b

Figure 2.8. Diagrams of pendulum tribometer (a) and electrochemical cell for polarizing friction zone (b): 1, tribometer; 2, pendulum; 3, support; 4, prism; 5, U-shape core; 6, sensor; 7, rectifier; 8, amplifier; 9, potentiometer; 10, stabilizer; 11, electrolyte; 12, auxiliary electrode; 13, working electrode; 14, electrolytic key; 15, vessel; 16, reference electrode.

In combination with other modern techniques, they enable better insights into the nature and mechanisms of friction of polymer materials.

2.5. DECORATION TECHNIQUE

The decoration techniques are widely applied for investigating the real structures of solids by decorating both the surfaces and the bulk, *ie* by selective crystallization of decorating substances on surface active points [33].

This technique can be also employed for visual representation of the electrical surface relief of polymers and its modification by mechanical and chemophysical effects. After decoration, the patterns are processed and analyzed to reflect the macrotopography of the surface active components.

The technique allows collection of data on the distribution of positive and negative charges on the surfaces of polymer electrets, visual representation of the dynamics of electrical processes evolving on the polymer surfaces, support of the electretization technique of investigating polymer electrets, optimization of the design of electrodes producing electrets by crown charges, investigation of the dynamics of emission processes, and investigatigation of friction of polymers. Decoration is achieved in the following manner. A specimen is secured inside a special vessel so that the carrier with adsorbed decorating particles can easily move over its surface. The suitable carrier is similar to the one used for electrophotography, and industrial carbon particles serve for decoration.

The positive sign of the charge carrier is established experimentally, so that carbon precipitates on negatively charged portions making the electrically active components visible. Optical systems produce photocopies of the patterns of decoration with magnifications of X10, X20, X50 or more.

Figure 2.9 shows the pattern of the decorated polyethylene surface with the optical systems providing a visual representation of the polymer surface before and after friction.

Figure 2.9.   Polyethylene decorated with industrial carbon (magnification X25). Dark background – regions with negative surface charge.

The edges of the picture show the crown electret before friction and the dendrite structure of surface charges, the center shows the pattern after friction (under the following conditions p = 0.1 MPa, V = -.8 m/s, dry friction). The electrical field effect upon the texture is smoothed out by friction: the space to the left of the unipolar friction path (the dark background) is the original dendrite structure, and the light background shows positively charged surface portions. Thus, it is possible to obtain the data on the distribution of positive and negative charges on the electret surface and a visual representation of the dynamics of electrical processes at friction.

Figure 2.10 shows the pattern of decoration of polyethylene electret surfaces   produced by the crown charge (a) or in the field (b). The differences between the structures of the two types of electrets are clearly visible: the dendrite structure of the crown charged electrets and the globular structure of the electrets produced in the field. Thus, the decoration patterns allow determination of the techniques employed to produce an electret, *ie* to identify the electretization technique.

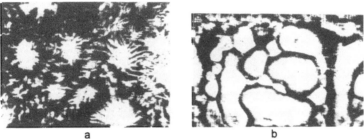

Figure 2.10. Polyethylene decorated with industrial carbon: a, crown electret; b, electret produced in field (magnification X10).

Thus, the technique allows more specific identification of the charge distribution over the electret surface and optimization of the design, of the charge producing electrodes for the crown discharge.
Figure 2.11 shows the patterns of surface decoration of a polyethylene electret produced by the crown charge from a plate electrode with needles.

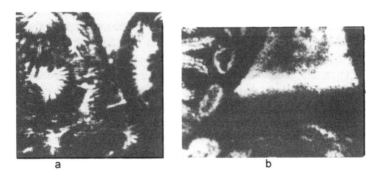

Figure 2.11. Polyethylene decorated with industrial carbon (magnification: a – X10; b – X50).

In Figure 2.11*a* clearly shows the pattern of charge distribution from individual needles (circles), the overlapping circles, and the "dead zone" region (in the center).The charge carrier missed the dead zone due to wrong selection of the pitch of the needles. Figure 2.11*b* shows a of the dead zone after strong magnification fragment.

Figure 2.12 shows the pattern of decoration of the friction path and the adjacent region after friction of PTFE on a metal.

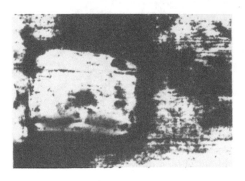

Figure 2.12. Polytetrafluorethylene decorated with industrial carbon after friction (magnification X2).

The charged regions outside the friction path are clearly visible, because the decoration of original specimens without friction would not produce any charged regions. The charged regions are apparently produced by the emission processes during friction of a polymer against a metal. The data obtained agree with the Tiessen magma-plasma contact model.

Thus, the decoration techniques yield a valid visual representation of the polymer surface electrical relief, which can be used for determination of the evenness of the distribution of charges and the distribution of positive and negative charges over the surface, investigation of the dynamics of the electrical processes in friction; and optimizing the design of charging devices, the process of electretization, etc. Video attachments make the process computerized and improve the information content.

REFERENCES

1. Bilik Sh.M. 'Metal-Plastic Friction Pairs in Machinery' (in Russian), Moscow, 1965.
2. Tsurkan, V.P. 'Plastics in Sliding Bearings' (in Russian), Moscow, 1965, p. 75.
3. Rogers, J.L. *SPE Journal* , 1973, **29**, no. 28, 52.
4. Shashoua, V.E. *J. Polymer Sci.* Part A, 1963, **1**, no.1, 169–187.

5. Patent registered by Bulgaria, N 31601, Device for Investigating Electrification, 1982.
6. Ohara, K., Uchigamas, S. and Takagi, H. *J. of Physics E: Scientific Instruments*, 1976, 9, 226–229.
7. Klimovich, A.F. 'Development and Investigation of Electrostatic Technique for Production of Fine-Layered Polymer Coatings' (in Russian), Ph.D. Dissertation, Gomel, 1970.
8. Gubkin, A.H. 'Electrets' (in Russian), Moscow, 1978.
9. Lushchejkin, A.G. 'Polymeric Electrets', (in Russian), Moscow, 1976.
10. Cessler, G. 'Electrets', (Russian translation), Moscow, 1983.
11. Collins, R.E. *Appl. Phys. Lett.* , 1975, 16, no. 3, 675–677.
12. De Regg, A.A. *Phys. Rev. Lett.* , 1978, 40 no. 6, 413–419.
13. Gorokhovatskii, Yu.A. 'Principles of Thermo-Polarized Analysis' (in Russian), Moscow, 1981.
14. Lushchejkin, G.A. 'Methods of Investigating the Electrical Behavior of Polymers' (in Russian), Moscow, 1988.
15. Turnhout, J. 'Thermally Stimulated Discharge of Polymers Electrets', N.Y., 1975.
16. Evdokimov, V.D. and Semov, Yu.I. 'Exoelectron Emission at Friction' (in Russian), Moscow, 1973.
17. Handsel-Poverta, Z., Pershkala, A. and Piruch M. *Soviet Journal of Friction and Wear*, Allerton Press, N.Y., 1981, 2, no.1, 15–18.
18. Shpenkov, G.V. 'Physicochemistry of Friction' (in Russian), Minsk, 1991.
19. Kramer, J. *Acta Phys. Austr.* , 1957, 10, no. 4, 327.
20. Deryagin, B.V., Krotova, N.A. and Khrustalev, Yu.A. in 'Active Surfaces of Solids' (in Russian), Moscow 1976, p.6.
21. Krotova, N.A. in 'Proc. of VII All-Union Symp. on Mechanoemission and Mechanochemistry of Solids' (in Russian), Tallinn, 1986, 58–67.
22. Karasev, V.V., Krotova, N.A. and Deryaguin, B.V. *Proc. of the USSR Acad. of Sci.* (in Russian), 1953, 89, 109.
23. Deryaguin, B.V. and Toporov, Yu.P. in 'Proc. of VII All-Union Symp. on Mechanoemission and Mechanochemics of Solids', part 1, Tashkent, 1981, 3–7.
24. Khrustalev, Yu.A. in 'Non-Equilibrium Processes in Dielectric Materials' (in Russian), Moscow, 1983, p. 47.
25. Sviridenok, A.I., Myshkin, N.K., Kalmykova, T.F. and Kholodilov, O.V. 'Acoustic and Electrical Methods in Triboengineering', Allerton Press, New York, 1989.
26. Lee, L.-H. in 'Adhesives and Adhesion Compounds' (Russian translation), Moscow, 1988, p. 4.
27. Goldade, V.A., Struk, V.A. and Pesetskii, S.S. 'Inhibitors of Wear of Metal-Polymer Systems' (in Russian), Moscow, 1993.
28. Nevzorov, V.V., Kirpichenko, Yu.E. and Sviridenok, A.I. *Soviet Journal of Friction and Wear*, Allerton Press, N.Y., 1983, 4, no.2, 25–29.
29. Ponomorenko, A.T., Shevchenko, V.G., Kryazhev, Y.G. and Kestelman, V.N. *Int.J. Polymer Mater.* , 1994, 25, 207–226.
30. Guzenkov, S.N. and Klimovich, A.F. *Journal of Friction and Wear*, Allerton Press, N.Y., 1992, 13, no.3, 52–56.
31. Preys, G.A. and Dzyub, L.G. *Soviet Journal of Friction and Wear*, Allerton Press, N.Y., 1980, 1, no.2, 18–31.

32. Pinchuk, L.S., Goldade, V.A. and Neverov, A.S. *Soviet Journal of Friction and Wear,* Allerton Press, N.Y., 1980, 1, no.4, 88–91.
33. Dastler, G.I., Vlasov, V.P. and Gerasimov, Yu.M. 'Decoration of Surfaces of Solids' (in Russian), Moscow, 1976.
34. Kelemen, K. and Pal, G. *Plaste und Kautschuk,* 1970, 12, 907.
35. Kao, K. and Huang, V. 'Electron Transfer in Solids' (Russian translation), Moscow, 1984.

CHAPTER 3

TRIBOELECTRIFICATION OF POLYMER POWDERS

Intensive mechanical and thermal effects during recycling and operation cause physical and chemical transformations in polymer powders accompanied by electrification, polarization, and electretization. A significant amount of research has been performed in this area.

3.1. STATIC ELECTRIFICATION OF LIQUID-CIRCULATED AND FREE SPRAYED POLYMERS

Systematic studies of electrification of moving powders have revealed that the number of positively charged particles in these materials approximately coincides with the number of negatively charged particles. They have also shown that the average charge of a particle increases with its diameter, that changes in the relative humidity up to 90% have no significant effect on the electrification process (except for strongly hydroscopic substances), that the asymmetric charging is observed when unlike materials contact, and that the pronounced asymmetry of charging has been observed when large and dense particles collide with some insulated surface or when a mixture of two unlike materials separates. The two processes may occur at low air humidity. The observed charging process is due to contact phenomena on surfaces [1-4].

A majority of studies of the static electrification of polymer powders under mechanical effects relates to the measurements of charge and size of individual particles. Powders are sprayed using special dozers in order to obtain a small concentration of airborne particles. Despite apparent advantages (a possibility to measure real charges and sizes of particles), these techniques have substantial drawbacks as they are labor-intensive and have poor reproducibility of results [5-7].

To investigate the processes of static electrification of polymer powders we developed a special technique and a special experimental device (Fig. 3.1).

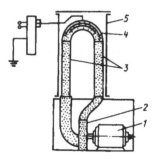

Fig. 3.1. Experimental set-up for investigation of the process of static electrification. 1, drive; 2, snail with impeller; 3, circulation pipes with dielectric (glass); 4, metallic trap; 5, frame.

It employs the principle of circulation of the airborne polymeric mixture in a closed cycle. The device consists of a drive, a snail with impeller, circulation pipes from dielectric (glass) and a metallic trap. This device is mounted on a metallic frame enclosed into a polymetylmethacrylate (PMMA) sheath. The variable drive allows variation of the velocity of polymeric particles within 0.5–20 m/s. Measurements of controllable generated potential and controllable short-circuit are performed by kilovoltmeters and a recorder, respectively.

When powder is confined to the closed cycle, the contact becomes dynamic because of repeated consecutive instantaneous contacts of particles with the walls of pipes and the trap. Active mechanical effects combining friction and collisions produce pronounced asymmetric electrification of the polymer powder. The space charge of the opposite sign accumulates in the trap. The closed cycle idea and the design of the device possess a valuable advantage so that large quantities of particles with strongly developed surfaces can be involved yielding rather high total charges. This facilitates measurements and improves the reproducibility of results.

Variations of shapes, sizes, and states of polymeric particles are monitored using an optical microscope. A special sampling probe was used to choose particles for the optical monitoring. It is a hard task to measure experimentally the local temperature of a microcontact, therefore a thermocouple was employed to monitor the heating temperature of the entire circulation system.

Experiments were performed with the following dispersed polymers: polyvinylbutyral (PVB), polycaproamide (PCA), high density polyethylene (HDPE), polypropylene (PP), polystyrene [(PS) emulsion-type], polytetrafluorethylene (PTFE), polyvinylchloride (PVC). Standard sieves were applied to size powders.

The results showed that the repeated contacts over the interposed polymer-metal boundary caused significant mechanical and thermal stresses resulting in the deformation of polymer particles as well as alteration of their shapes and sizes. Collisions and melting during circulation transform the loose irregularly-shaped PCA particles into dense semitransparent rounded particles up to 300 μm in diameter and transparent fibers more than 600 μm long.

From the friction conditions, the temperature in the contact zone T [3] is given by

$$T = T_0(A \cdot \frac{\sqrt{\pi}}{2} - 1), \quad A = \frac{F\mu v\sqrt{at}}{kT_0}, \quad (3.1)$$

where T_0 is the melting point of the material; F is the force with which particles are pressed against the surface, μ is the friction coefficient between the particles and the surface, v is the velocity of the particles relative to the surface, k is the heat conductivity or diffusivity of particles, and t is the duration of contact.

Melting does not occur at $A < A_{cr} = \frac{\sqrt{\pi}}{2}$, since then $T < T_0$. Yet, the estimates using equations (3.1) have revealed that melting is unachievable even when the entire work of friction forces convert into heat. Still, visual observation indicates that melting does occur. Conversion of collision energy into heat is apparently a governing factor in addition to friction. Local microcontact temperatures may reach 10^2–10^3 K, taking into account the experimental factors of melting of large polymer particles with the melting point about 473 K.

Also, the theory of boundary friction of metals states that local microcontact temperatures may exceed 1000 K generating heat that may transform boundary layers into both liquid and gaseous states. Moreover, plasma states may occur during collisions in microcontacts. Bowden and Tabor have specifically noted that microscopic volumes of a material in friction can undergo phase transitions from the solid state into the plasma states, possibly bypassing intermediate stages (liquid and gaseous states).

According to Frenkel [8], the closer the collision time to the period of oscillations of internal particles of a body, the more the body behaves like a solid.

It should be noted that the oscillation period of atoms and ions near their equilibrium in solid crystalline bodies is $\approx 10^{-13}$ s, the time of the collisions at moderate velocities is about $10^{-6}-10^{-7}$ s, or even $10^{-13}-10^{-12}$ s.

Meanwhile, even thoroughly polished specimens have roughness with sizes up to 100Å, which creates significant area of the physical contact. High temperature and pressure at these sites will cause the material to deform.

The processes in the thin surface layer (adhesion, elastic and plastic deformations, structural and plastic transformations) should be treated as high-speed processes as their kinetics change within the time frame of the short, periodic elementary contacts. This results in high local temperatures.

High microcontact temperatures favor transition of charge carriers and more intensive electrification for various reasons. First, higher temperatures strongly reduce the specific resistance of dielectrics. Second, the transfer of charges relates to the thermal motion of both the whole macromolecule and individual radicals or monomer chains. Whereas this process relates to local modifications of the conformation of small fragments of the macrochain in the glassy state, it is caused by the significant motion of the macrochain in the highly elastic state. Third, melting of polymer particles favors their agglomeration, makes them homogeneous and more compact. Comparison of the mass of the fresh PV and one after 100 hours in the circulation system has manifested a significant increase in mass of the mechanically activated powder (by 50–55%) (Table 3.1).

Table 3.1. Bulk mass of original polyvinylbutyral and the one after 100 h of circulation

Polymer condition	Particle size, μm	Bulk mass, g/cm³
Original	100–160	0.32
After 100 h of circulation	100–160	0.49
After 100 h of circulation	Unfractioned	0.50

In its turn, further cycles increase the energy of collisions, thus intensifying the powder conversion into the plastic state and reducing the "clearance" along the polymer - metal boundary. Appearance of fibers, spirals, strips and films (Fig. 3.2) in the process of powder circulation demonstrates both the melting in the contact zone and the seizure of contacts. The transfer of carriers and the electrification rate i intensify.

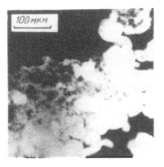

Fig. 3.2. Transformation of polyethylene powder after 4 h in circulation system.

Studies have revealed that the repeated collisions of the polymer particles with the metallic trap walls cause their intensive asymmetric electrification. The electrostatic potential in the trap may reach 200–300 kV.

The rate of electrification, and, hence, the magnitude of the voltage generated is a function of the charge production and attenuation on polymer particles. The rate of appearance and attenuation of the charge, in their turn, depends on the characteristics of contacting surfaces (metal – dielectric, in our case); the closeness of the contact; and, especially, upon the rate of contact cycles, *ie*, the velocity of polymer particles. As the velocity goes up, the electrification intensity increases (Fig. 3.3).

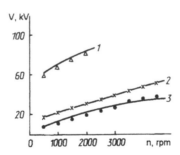

Figure 3.3. Trap electrostatic potential vs. velocity of circulating particles (impeller rpm.) and external factors for polyvinylbutyral; powder concentration is 2.5 g/l; particle size is 100–200 µm; relative air humidity: 1, 56%; 2, 65.3%; 3, 72%)

The increase in the electrostatic potential of the trap observed when the velocity of particles grows is caused both by the greater frequency of contact repetitions and by the smaller clearance between the contacting bodies or between the polymer particles and the trap due to stronger impacts.

Table 3.2 shows the experimental results illustrating how the electrostatic potential of the trap depends on the concentration of particles in the circulating mixture. It is apparent that the significant changes in concentration only moderately affect the electrification process (within certain limits).

Table 3.2. Trap electrical potential U, kV versus concentration of circulating particles, g/l (polyvinylbutyral, particle size 100–200 μm, air humidity 54...58%).

Concentration of particles, g/l	Velocity of particles (impeller n, rpm)				
	500	1000	2000	3000	4000
0.8	41	52	62	2	10
1.7	25	55	80	90	–
2.5	9	22	50	76	–
3.5	48	70	80	84	–
4.2	64	69	72	79	–
5.0	47	60	70	75	–
6.0	32	31	30	48	90
7.0	38	57	68	73	–
8.5	–	4	22	10	13
17.0	–	1	3	4	2

When the concentration of particles is $z \leq 1$ g/l (the lower limit), initially, the electrostatic potential goes up, but later it drops causing the unstable operation. A sharp drop of the potential is due to the adhesion of the polymer to the circulation system pipes and depletion of the mixture. When the concentration is $z \geq 8$ g/l (the upper limit) a poor electrification intensity is due to the changes in the circulating flow hydrodynamics. The optimum concentration is about 1–7 g/l.

The morphological studies have showed that circulation produces both large and fine fractions. The size of the fine fraction is independent of the original size of the particles, as they gradually approach the final size (about 10–20 μm in the experiment).

The particle size of the large fraction depends on the initial size and the polymer type: PCA and PP yield large round- or even square-shaped particles up to 300–400 μm. Polyethylene (PE) during "protracted" circulation covers the trap with transparent films (Fig. 3.2) without yielding any larger particles.

The sieve analysis demonstrates the domination of fraction size coinciding with the original size (Table 3.3).

Table 3.3. Polycaproamide fraction sizing (original particle size 100–160 µm, duration of mechanical activation 8 h).

Particle size, µm	Percentage, %	Notes
71	5	Pulverization
71–100	29	Pulverization
100–160	66	No pulverization

It can be assumed that the electrification intensity declines gradually if it is caused by pulverization only. Yet, the experimental studies of the electrification intensity after "protracted" circulation prove the opposite (Fig. 3.4) – the effect intensifies.

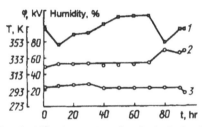

Figure 3.4.   Static electrification rate of polyvinylbutyral during protracted circulation (concentration 5g/l; original size 100–160 µm): 1, electrostatic potential; 2, relative air humidity; 3, temperature.

Studies of the electrification intensity as the function of the nature of polymer particles (Table 3.4) show that polar polymers get electrified quicker than nonpolar ones.

Table 3.4. Electrification rate as the function of the polymer nature and duration of mechanical activation, hr (particles concentration 3–5 g/l; particle size 100–200 µm; impeller rpm 150; relative air humidity 50–60%).

Material	Trap voltage, kV					
	Start	1	2	3	4	5
Polyvinylbutyral	81	90	80	85	80	77
Polycaproamide	70	71	70	68	65	62
Polyvinylchloride	50	55	62	58	56	60
Polypropylene	2	22	45	60	58	55
Polyethylene	2	10	12	15	28	30
Polystyrene	2	15	22	28	30	25
Polytetrafluorethylene	60	52	35	22	10	–

A relation can be found between the existence and the reactivity of functional groups governing the molecular polarity and the electrification intensity.

Polyvinylbutyral whose structural formula is given by

$$\ldots\text{---}CH_2\text{---}CH\text{---}CH_2\text{---}CH\text{---}\ldots$$

with O and O bridging to CH—C_3H_7

contains 65–78% of PVB, 32–19% of polyvinyl alcohol and about 3% of polyvinyl acetate in its macromolecule:

$$\ldots\text{---}CH_2\text{---}CH\text{---}\ldots \text{ (polyvinyl}$$
$$\qquad\qquad |$$
$$\qquad\qquad OH \qquad\qquad \text{alcohol);}$$

$$\ldots\text{---}CH_2\text{---}CH\text{---}\ldots \text{ (polyvinyl}$$
$$\qquad\qquad |$$
$$\qquad\qquad OCOCH_3 \qquad \text{acetate).}$$

The high concentration of polar butyral, hydroxyl and acetyl groups $(OCOCH_3)$ favors strong polarization of PVB and leads to intensive triboelectrification.

Polycaproamide with the structural formula

$$\ldots\text{---}(CH_2)_5\text{---}C\text{---}N\text{---}\ldots$$

with H on N and $\parallel O$ on C

contains a great number of polar $\text{---}C\!\!\begin{smallmatrix}\nearrow NH\text{---}\\ \searrow O\end{smallmatrix}$ groups. Intra- and intermolecular hydrogen bonds typical for PCA make this compound quickly polarizable.

Polyvinylchloride $\ldots\text{---}CH_2\text{---}CH\text{---}\ldots$ with Cl contains a great number of strongly negative and reactive chlorine ions determining the polymer polarity. That is why PVA and PVC are easily electrified.

Polytetrafluorethylene has significant dielectric losses due to the asymmetry of the main molecular chain. Theoretically, PE and PP are nonpolar polymers. Yet, they have significant dielectric losses caused by the presence of a small concentration of the nonpolar groups forming impurities in the polymer during the fabrication or the operation. Thus, industrial PE contains small concentrations of hydroxyl, carboxyl, carbonyl and other polar groups. A typical noticeable feature of the carbonyl group is that the carbon atom has a deficit of electrons, *ie* it has a partial positive charge. As far as their main chain is concerned, the above functional groups and atoms may act as electron donors (nucleophils, donating an electron) or electron acceptors (electrophils) or reagents, whereas strongly polar groups perform both functions. The functional groups with electron donating and electron attractive behavior can be arranged in the order of declining reactivity:

electron donors:

—OH, OCOCH$_3$, OR (butyral), CH$_3$, F, Cl, C$_6$H$_5$;

electron acceptors:

$$-NO_2 \left(-N\diagup^{OH}_{\diagdown O}\right), \quad CO\left(\!\diagup C=O\right), \quad -C\diagup^{NH-}_{\diagdown O}\;.$$

Thus, the electron donor groups possess an excessive negative charge, the electron-attracting groups show the opposite behavior. Out of the above polymers PF, PTFE, PVC, PP, and PC contain electron donor groups, PCA - electrophilic groups. Thus, it can be concluded that the intensity of the electrification depends only slightly on the electron donating or electron attracting behavior of the functional groups, it is primarily determined by their reactivity during the recycling and the operation .

Tables 3.5–3.7 and Figures 3.5, 3.6 show the effect of the nature of a circulating polymeron on the quantity and polarity of the short circuit current (SCC) flowing from the trap and upon the created electrostatic potential (the electrification intensity).

Table 3.5. Effect of trap material upon electrostatic polarity and short circuit current polarity (relative air humidity 43–49%).

Material of circulating particles	Trap potential, kV		Potential polarity	
	Brass	Aluminum	Brass	Aluminum
Polyvinylbutyral	58	60	+	+
Polycaproamide	12	10	+	+
Polyvinylchloride	15	12	+	+
Polypropylene	62	13	+	+
Polyethylene	65	40	-	-
Polystyrene	64	45	+	+
Polytetrafluorethylene	8	10	-	-

Table 3.6. Short circuit current magnitude and polarity versus nature of circulating polymers.

Material	Short circuit current magnitude and polarity at various circulation duration (hr), μA			
	2	4	6	8
Polyvinylbutyral	+4.0	+3.5	+3.6	+6.0
Polycaproamide	+6.0	+0.7	+6.0	–
Polyvinylchloride	–2.0	–4.0	–4.7	–2.5
Polytetrafluorethylene	–2.0	0	+0.3	+0.8
High density polyethylene	–0.7	+1.5	+1.7	+3.0
Polypropylene	+1.5	+1.5	+3.0	+4.5
Polystyrene	–6.0	–6.3	–6.0	–4.8

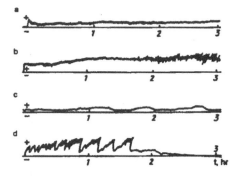

Figure 3.5. Short circuit current plots of polar polymers: a, polyvinylbutyral; b, polytetrafluorethylene; c, polyvinylchloride; d, polycaproamide.

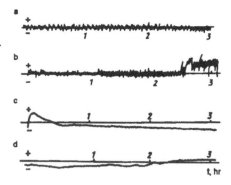

Figure 3.6. short circuit current plots of nonpolar polymers: a, b, polyethylene; c, polypropylene; d, polystyren

Table 3.7. Electrostatic potential (V) and short circuit current (J) versus polymer powders percentage in circulating mixture (relative air humidity 41–50%).

Components	concentration, %	V, kV	Duration of circulation, hr			
		J, µA	2	4	6	8
Polycaproamide (PCA)	85	V	40	62	65	60
Polyvinylchloride (PVC)	15	J	+6.0	+5.5	+6.0	+3.8
PCA	70	V	80	78	46	70
PVC	30	J	+4.5	−8.0	+9.0	+10.6
PCA	50	V	5	3	5	6.5
PVC	50	J	+2.7	+3.5	+4.8	+4.3
PCA	30	V	75	75	75	75
PVC	70	J	–	+7.0	+9.0	+9.0
PCA	15	V	60	74	74	70
PVC	85	J	+6.0	+12.5	+7.5	+8.0
PCA	85	V	24	17	15	15
Polystyrene (PS)	15	J	+0.07	+0.09	+0.08	+0.07
PCA	70	V	36	45	55	55
PS	30	J	–	+0.15	+0.10	+0.09
PCA	50	V	8	15	14	15
PS	50	J	+2.4	+5.0	+3.6	+3.5
PCA	30	V	3	3	5	5
PS	70	J	−2.4	+0.15	+0.93	+1.70
PCA	15	V	9	6	7	5
PS	85	J	−3.0	−7.0	−2.0	−1.4
Polyvinylbutyral (PVB)	85	V	6	7	7	6
PS	15	J	+4.0	+5.4	+3.7	+3.6
PVB	70	V	4	5.5	6	6
PS	30	J	+4.0	+4.8	+3.7	+3.0
PVB	50	V	0	0	3	0
PS	50	J	–	–	–	–
PVB	30	V	0	2	1	0
PS	70	J	−1.1	−0.9	−1.1	−1.0
PVB	15	V	1	2	1	1
PS	85	J	−1.5	−1.4	−0.9	−1.0

Brass trap, body and fan impeller were used in all the tests, except the studies of the effect of the trap material.

Experimental results show that the static electrification of polymer powders during spraying or circulation in a stream (the dynamic contact) is characterized by high total charges (the potential of the insulated trap reaches 100 kV and more). They also show that the static electrification of dispersed polymers is characterized by the current inversion (the majority of polymers show this after mechanical activation during 3 hours), the charge polarity of powders generated in contact with the metallic surface is governed by the nature of a polymer and its behavior, and that the dominating polarity is typical negative for polar dielectrics (PVB, CPA, PVC and others), ie for the polymers with larger dielectric constants. They usually become negatively charged when in contact with metals. A more stable charge polarity is typical for polar dielectrics. Electron-donating or electron-attracting behavior of the functional groups have no effect on the charge polarity. It is identical for the PCA, containing electron-attracting groups, and for the PF containing electron-donating groups. Substitution of the brass trap with an aluminum trap has no effect on the charge polarity of all the polymers studied. The electrification intensity and the SCC polarity of a mixture of two dielectrics depend on the ratio between the components. The electrification intensity of a mixture of two polar dielectrics is governed by the dielectric dominating in the mixture; electrification is inhibited when the concentration of the components are equal. The polar dielectric determines the electrification intensity in a mixture with a nonpolar dielectric.

Equal proportions of components or smaller concentration of the polar component can completely suppress electrification and yield zero potentials. The SCC polarity in a mixture of two polymers (PCA–PS, PVB–PC) with opposite polarities is governed by the component dominating in the mixture.

The results of the above experimental studies have demonstrated practical applicability of the static electrification of polymer powders in designing the equipment for deposition of polymer coatings.

The spraying of polymer powders frequently leads to the problem of their electrification [6]. Powders with particle size under 30 µm usually demonstrate asymmetric charging. As the size of particles increases, so does the asymmetry in the charge distribution. Usually powders with particles of 100–300 µm are used for polymer coatings. It is interesting to trace the electrification of such particles and to compare the results to those produced by smaller particles.

Figure 3.7 a shows the design of the experimental set-up for studies of these effects. The airborne powder is sprayed with a gun between the plates of a slot capacitor held under high voltage. The spraying gun has coil-like inner passages to increase the number of collisions between the particles and between the particles and the walls. All parts are made from PMMA, and the walls are lined with a PE film. The following polymers were tested: PE, PP, PVB and PCA produced by repeated precipitation

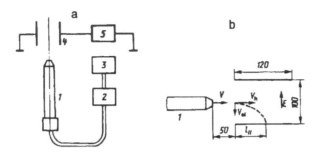

Figure 3.7. Experimental setup (a), view from the top on slot capacitor plates and velocity vectors of particles (b); 1, spray gun; 2, powder spraying unit; 3, compressor; 4, slot capacitor; 5, high voltage source.

from solution with ε-caprolactam. The size of particles in the tests was 50–315 µm.

As a characteristic of the charging unipolarity of polymer particles, the following quantity has been selected:

$$\alpha = \frac{P_-}{P_- + P_+} \cdot 100\%. \qquad (3.2)$$

Here P_- is the powder mass deposited on the positively charged plate, ie the mass of negatively charged particles and P_+ is the positively charged powder mass.

The mode of spraying of the airborne powder mixture was selected based on the estimate of motion of the powdered PP in the slot capacitor field for the particles in three charge states: 1) maximum charge, 2) charge two orders of magnitude less than the maximum, and 3) zero total charge [10].

The maximum charge of a PE particle of 100 µm in diameter is $Q_{max} = 4\pi\varepsilon_0\varepsilon E_0 = 3.75 \ 10^{-12}$ C, where $E_0 = 30$ kV/cm is the sparkover voltage for the air; r is the particle radius; ε_0 is the dielectric constant; and ε is the dielectric permeability of the particle material.

To estimate the motion of the charged particle in the slot capacitor field, it is necessary to know its trajectory after it escapes from the gun. Motion of a single particle with mass m, density ρ_1, and the initial velocity v_{in} in a static gas medium with the density ρ is described by the equations [11]:

$$m\frac{dv_h}{dt} = \frac{cfpv^2}{2}\cos\alpha,$$ (3.3)

$$m\frac{dv_v}{dt} = mg\frac{\rho_1-\rho}{\rho} - \frac{cfV^2}{2}\sin\alpha,$$ (3.4)

where m is the particle mass ($m = \pi D^3 \rho_1/6$ for a spherical particle with the diameter D); v is the velocity of the particle; v_h and v_v are horizontal and vertical velocity components, respectively ($\cos\alpha = v_h/v$; $\sin\alpha = v_v/v$); t is time; f is the middle cross-section area (for a spherical particle $f = \pi D^2/4$); and C is the resistance factor depending on the Reynolds number Re; g is gravity acceleration.

For a spherical particle with the velocity vector at the moment of escape from the gun coinciding with the horizontal axis of the slot capacitor, the velocity components v_h and v_v are determined by integrating equations (3.3) and (3.4):

$$v_h = ve^{-\frac{3c\rho V}{4L\rho_1}t},$$ (3.5)

$$v_v = \frac{4}{3}\frac{Dg(\rho_1-\rho)}{c\rho v}\left(1 - e^{-\frac{3c\rho v}{4D\rho_1}t}\right).$$ (3.6)

Here v is the particle velocity at the moment when it escapes from the gun, which can be determined from the delivery pipe diameter and the flow rate.

For the above parameters of the particles it has been established that $v=1.76$ m/s. The resistance factor c in the considered case is equal to 0.4 (it depends on the shape of the particle and the Reynolds number). Using expressions (3.5) and (3.6) and the known values of the parameters the following equations have been derived for determining the trajectory of the particle after it has escaped from the gun:

$$v_h = 1.76\, e^{-13.6t},$$ (3.7)

$$v_v = 7(1 - e^{-13.6t}),$$ (3.8)

Charged particles, once they penetrate into the region between the plates of the slot capacitor, are diverged by the field of one of the plates (Fig. 3.7 b) according to the charge sign (the component v_{el}). To determine the trajectory of a charged particle in the capacitor, the period of time t_C since the moment it gets into the capacitor until it precipitates on the plate has to be determined. It can be found from the formula

$$l_\perp = \int_0^{t_c} V(t)dt = \int_0^{t_c}\int_0^{t_c}\frac{F_e}{m}dt^2 = \frac{3gFt_c^2}{\pi D^3 \rho_1},$$ (3.9)

where $l_\perp = 5$ cm is the distance from the central axis to the capacitor plate; F_e is the electrical force. Hence,

$$t_c = \sqrt{\frac{\pi D^3 \rho_1 l_\perp}{3gE}}.$$ (3.10)

During time t_c the particle travels horizontally from the point where it gets into the capacitor to the place where it precipitates on the plate:

$$l_\Pi = \int_0^{t_c} v_b(t)dt = \frac{4D\rho_1}{3cp}\left(1 - e^{-\frac{3cp v_e}{4D\rho_1}t_e}\right).$$ (3.11)

Here v_0 is the velocity of the particle at the moment when it gets into the capacitor.

The estimate with the help of the above formulae has revealed (Table 3.8) that charged polymer particles precipitate on the plates under the effect of the field, whereas neutral particles precipitate on the bottom of the capacitor or outside of it .

Table 3.8. Parameters of polymeric particles motion in slot capacitor field.

| D, μm | Charged particles | | | | Neutral |
	Q_{max}, C	v_0, m/s	t_c, s	$l_{\|\|}$, cm	particles l_x, cm
50	$0.48 \cdot 10^{-12}$	0.65	$0.8 \cdot 10^{-2}$	0.52	6.5
100	$1.91 \cdot 10^{-12}$	1.2	$1.12 \cdot 10^{-2}$	1.3	13
200	$7.65 \cdot 10^{-12}$	1.53	$1.6 \cdot 10^{-2}$	2.3	26

Note. l_x is the distance from the gun to the bottom of the capacitor .

Figure 3.8 shows the plot of the factor α versus the time of spraying of various polymers (the particle sizes are 100–160 μm) at the voltage of 1 kV/cm in the slot capacitor. When the spraying time of nonpolar polymers (PE and PP) is extended, the factor α increases for the time span studied.

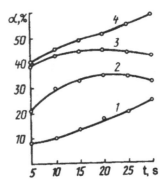

Figure 3.8. Factor α vs. duration of polymer powders spraying: *1*, high density polyethylene; *2*, polyvinylbutyral; *3*, polycaproamide; *4*, polypropylene.

For example, the factor α of polyethylene (curve *1*) increases by more than two times once the spraying time is extended from 5 to 30 s.

Polar polymers are a different case. Initially, the factor α of PVB and PCA follows the extending spraying time and later starts to reduce. Similar results have been obtained with the field intensity in the slot capacitor being 2.5 kV/cm. It is hard to draw any comparison at higher field intensities (3 kV/cm). In this case, the field of the capacitor, in the first place, affects the experimental results, *ie* the precipitation of particles on one or the other plate of the capacitor. The result is that at a certain moment the reverse crown charge appears that blows the powder off the plate once a layer of certain thickness has precipitated. Thus, the recharging is possible in a strong field causing variations in α.

Figure 3.9 shows the relation between this factor and the degree of dispersion of the sprayed PE. Various PE fractions (50–100, 100–160, 160–200, 200–315 μm) were sprayed during 15 s with the field intensity in the slot capacitor equal to 1 kV/cm. It has turned out that α increases as the particles become smaller, *ie* as the relative quantity of negatively charged PE particles increases.

We assume that at contact electrification of polymers, the charge carriers are electrons. Because they are more mobile than ions the energy of bonds is smaller. The particles of a dispersed polymer in contact with PE film give away electrons, *ie* they get charged positively. Such division of charges occurs because the dispersed polymer has a much better developed effective surface than the contacting film.

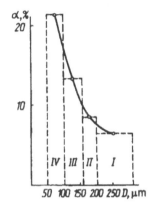

Figure 3.9. Factor α vs. dispersion of sprayed polyethylene.

Therefore, the particles have a much greater quantity of impurities on their surfaces, which increases the number of energy levels and allowed states for electrons leading to the weakening of bonds. Only if the effect of walls is taken into consideration, can symmetric charging be achieved, *ie* the numbers of particles charged positively and negatively coincide ($\alpha \sim$ 50%). With the effect of walls taken into account, the number of positively charged particles will be greater as polymer particles in contact with the walls would release electrons. Hence, the factor α will be less than 50%. As the time of spraying is extended, the probability of electron transfer will increase because the negatively charged film would repulse electrons. Therefore, the relative quantity of positively charged particles in the stream reduces meaning that α increases.

Now, let us consider the behavior of various polymer fractions in spraying. The greater the effective area of contact between a particle and a wall, the higher is the effectiveness of electrification of polymer particles during their friction against the walls. Yet, the larger the particles, *ie* the less their surface curvature is, the greater is the effective contact area leading to the smaller probability of electrification. Moreover, the probability that larger particles in the stream would collide with the walls is higher than the same probability for smaller particles. Hence, there are more chances that larger particles would release electrons than small particles, *ie* the relative quantity of positively charged polymer particles would increase as the size of powder particles increases, and the reverse is true for the negatively charged particles.

Thus, the following conclusions can be made (1) as the spraying time of polymer powders extends, the relative quantity of negatively charged particles increases;   material with low specific resistance should be selected in order to achieve a flow of particles with a constant charge state during spraying (2) asymmetric unipolar charging occurs when polymeric powders with particle sizes 50–300 μm are sprayed, this asymmetry increasing with the particle size.

The results of the studies of the dispersion electrification are broadly applied for developing electrostatic techniques of fabricating fine polymeric coatings and for other manufacturing processes.

3.2. VARIATIONS IN STRUCTURES AND BEHAVIOR OF POLYMERS

Studies of dispersed polymers establish that cyclic processes in repeated polymer-metal contacts cause significant deformation of polymer particles and their structural changes [5,6,12,13]. The structural and chemophysical transformations of the particles have been studied using structural radiographic, differential thermal (DTA), thermogravimetric (TGA) analyses, and infrared spectroscopy. Radiographic diffraction plots were recorded at the goniometer sweeping speed of 2 deg/min and with exposure to Cu_α ($\lambda_{Cu\ \alpha}$ = 1.542 Å). Derivatograms were registered in the air using an OD–102 apparatus with the heating rate of 5 K/min. IR-spectra were registered for identical weighed samples using a double beam spectrometer UR-20 within the bands of 400–4000 cm^{-1}. Variations of the masses of polymer powders at fixed temperature were measured. Also, the electric permeability of granulated polymers and films made from original mechanically activated powders was evaluated.

The process of dynamic contact between polymer particles and the trap described above includes impacts and sliding or rolling friction with slippage. When polymer particles travel with a speed of about 10 m/s they may collide with the trap 7 or 8 times per second. A significant portion of collision energy, as well as friction, is released as heat accumulated by the trap, pipes, and particles, and, partially dissipated into the surrounding space. Due to high temperatures of the microcontact, the particles make transition into the a plastic state (partially or fully) already in the initial period of circulation. When particles leave the contact zone they become airborne for about 0.1 s until the next collision. Part of the accumulated heat is dissipated into the space causing partial crystallization of the material. Thus, in the process of circulation, the polymeric particles periodically experience heat impacts and melting – crystallization phase transition, ie they undergo an intense mechanical changes accompanied by restructuring and modification of chemophysical behavior, as demonstrated by thermographic studies [7,19,21].

Figures 3.10–3.13 show the thermographic curves of the powdered PVB, PP, PS, and PTFE before and after activation by friction. The heating thermograph of the original amorphous PVB (Fig. 3.10 a, curve 1) is characterized by two endothermal peaks at 339 and 517 K and two exothermal peaks at 439 and 538 K. The endothermal peak at 339 K is due to the water desorption process, the exothermal peaks are due to the processes of oxidation of the polymer by the atmospheric oxygen and by thermoxidizing destruction. The thermographs of the mechanically activated PVB (Fig. 3.10, curves 2, 3) show that the peaks corresponding to the water desorption and thermooxidation appear at lower temperatures, ie they are shifted to the left. At this thermogram, the area under the exothermic curves at 423 and 503 K significantly exceeds the area under the similar peaks at 439 and 538 K of the original PVB. It indicates that the former is more responsive to the oxidation. As frictional activation increases (Fig. 3.10) the endothermal peak with the maximum at 338 K shifts towards smaller temperatures due to the activation energy drop in the process of water desorption from the polymer.

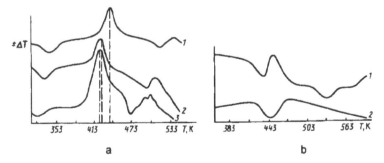

Figure 3.10. Thermograms of dispersed polyvinylbutyral (1) before thermal activation; (2) after 28.6 ks; (3) after 115.2 ks. (b) Thermograms of dispersed polytetrafluorethylene (1) after 90 ks of thermal activation; (2) before thermal activation.

Based on the data of infra-red (IR) spectrometry, we can say that dynamic contacts between dispersed PVB and the metal lead to a higher concentration of oxygen-containing groups C–OH and the appearance of cyclic structures in the polymer evidenced by a stronger absorption within the frequency range of 950–1200 and 1240–1250 cm^{-1} [14]. A stronger reactivity of the mechanically activated PVB in thermooxidation reactions is also confirmed by the reduction of the activation energy of the thermooxidation process determined according to the technique of Piloyan [15]. The estimated activation energy E_a in the original PVB amounts to 5.77 kcal/mol, and it becomes   4.07 kcal/mol after mechanical activation during 28.8 ks.

The thermograms of the chemically inert PTFE (Fig. 3.10 b) before and after mechanical activation indicate that dynamic contacts with the metal produce a significant structural transformation of the polymer accompanied by its partial amorphous state, and the reactivity of the polymer greatly increases its thermal oxidation. This is evidenced by the reduction of the endothermal peak at 446 K, which shifts towards 435 K and by the appearance of the exothermal peak at 453 K with its area growing as mechanical activation continues.

The mechanically activated PS powder (Fig. 3.11 a, curve 2) has the exothermal peak appearing on the graph at 435 K, ie it shifts strongly towards lower temperatures. The exothermal peak at 435 K after dynamic

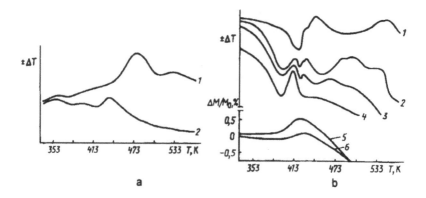

Figure 3.11. (a) Thermograms of polystyrene powder (1) before and (2) after mechanical activation during 32.4 ks. (b) Thermograms of dispersed polypropylene (1) before and (2) after mechanical activation 9 ks; (3) 18 ks; (4) 72 ks. Thermogravimetric curves of dispersed polypropylene (6) before and (5) after mechanical activation during 72 ks.

contacts of the dispersed PS with the metal is due to the process of the chemomechanical destruction accompanied by the appearance and stabilization of free macroradicals and accumulation of the low molecular mass products.

The graph for the original PP powder (Fig. 3.11 *b*) shows the peak at 418 K corresponding to the melting of the crystalline phase, and two exothermal peaks at 443 and 538 K that may be attributed to the process of oxidation of isostatic polypropylene. The graph for the PP powder after mechanical activation for 9 ks (Fig. 3.11 *b*, curve *2*) shows two endothermal peaks at 398 and 418 K and three exothermal peaks at 413, 430, and 483–497 K, respectively. It looks analogous to the graph of thermal oxidation of the atactic PP. The graph of PP activated for 18 ks (Fig. 3.11 *b*, curve *3*) shows that the maxima of the peaks at 430 and 483–497 K are shifted towards lower temperatures and their intensities are much less. As mechanical activation extends beyond 72 ks, the graph changes: the endothermal peak at 418 K and exothermal peak at 430 K completely disappear, whereas the intensity of the exothermal peak at 413 K significantly increases (Fig. 3.11 *b*, curve *4*).

The shift of the endothermal melting peak from 418 to 493 K apparently evidences the origination of low molecular mass products of mechanodestruction in the bulk of the polymer, the blurring of the peak, and the reduction in the area below is indicating the reduced concentration of the crystalline phase. A sharp exothermal effect on the graph is apparently due to a strong reactivity of the mechanically activated PP in thermal oxidation reactions. This is also shown by the data of the thermogravimetric analysis (Fig. 3.11 *b*, curves *5*, *6*). The mass of the mechanically activated PP specimen increases due to the oxidation reaction in the region of the exothermal peak maximum at 413 K.

The kinetics of variations of the mass of the dispersed PP at the temperature corresponding to the maximum of the considered thermal peak, *ie* at 413 K, have been studied to obtain additional information about the nature of the exothermal peaks. It follows from Table 3.9 that the exothermal peak of the mechanically activated PP is due to the processes of oxidation by atmospheric oxygen and the desorption of low molecular mass products appearing during mechanical destruction. The desorption process dominates over the oxidation processes. This is proven by large mass losses of the mechanically activated PP compared to the original mass during the initial period of time ($m_2/m_1 = 15$ at $t = 1.9$ ks). The observed reduction of the m_2/m_1 ratio proves the appearance of the mechanodestruction and desorption of low molecular mass products, with the increase in the contribution of thermodestruction as the exposure increases.

Table 3.9. Mass loss kinetics parameter $m = \Delta M/M_0$, % of dispersed polypropylene before and after mechanical activation under the exposure to atmosphere in thermostat at 413 K.

Polypropylene dispersion	Exposure time, ks						
	1.8	3.6	7.2	10.8	14.4	21.6	93.6
Original, m_1	0.08	0.10	0.40	0.70	1.10	1.80	12.1
After mechanical activation during 72 ks, m_2	1.20	1.30	1.70	2.10	2.80	3.40	12.1
Ratio m_2/m_1	15.0	13.0	4.2	3.0	2.5	1.9	1.0

The IR-spectroscopy corroborates the results of the thermal analysis. Figure 3.12 represents the IR-spectra of films with the same thickness made from original and mechanically activated powders by hot compassion at 473 K. The IR-spectrum of the mechanically activated PP powder (Fig. 3.12, curve 2) shows the increasing intensity of the absorption bands 1170, 1110, 975, 845, and 810 cm^{-1} verifying the increasing concentration of CH groups. The increasing absorption in the region 1640–1680 cm^{-1} evidences the increasing concentration of carboxyl groups C=O combined with aliphatic bonds C=C [16,32]. It should be noted that the stronger absorption within 1550–1560 cm^{-1} proves apparently the origination of metal-containing compounds like salts of fatty acids in the mechanically activated PP.

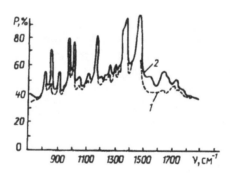

Figure 3.12. Infra-red spectra ($\sigma = 100$ µm) of films made of dispersed polypropylene (1) before mechanical activation and (2) after 72 ks of activation.

According to the radiographic structural analysis,   the degree of crystallinity of the studied polymers reduces by 5–10%. The maxima of the diffraction bands of the mechanically activated PP (Fig. 3.13, curve 2) shift by 12-15 min towards larger reflection angles .

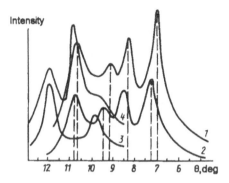

Figure 3.13. (1,2) Radiographic diffraction plots of dispersed polypropylene and (3,4) polycaproamide (1,3) before and (2) after mechanical activation during 72 ks; (4) after 108 ks .

The estimate indicates (Table 3.10) the deformation of crystals and reduction of interplanar distances in the crystalline lattice by 0.1–0.2 Å.

Table 3.10. Results of processing of radiographic diffraction plots of pelletized polypropylene (PP) powders.

Specimen	Reflection planes	Reflection angle	Intensity of lines based on 100 scale	Measured line width, mm	Crystallite size, Å	Interplanar distance, Å
Original PP	(110)	7°00'	20.0	13	77.72	6.324
	(040)	8°24'	15.7	10	105.24	5.277
	(130)	9°18'	9.0	12	85.49	4.771
	(111)	10°39'	12.8	19	52.29	4.170
PP after	(110)	7°15'	15.0	12	85.05	6.109
mechanical	(040)	8°39'	12.0	11	94.17	5.126
activation	(130)	9°30'	10.6	20	49.36	4.673
during 72 ks	(111)	10°51'	7.5	11	94.78	4.114

Figure 3.13 shows the radiographic diffraction plots of original (3) and mechanically activated (4) pelletized PCA powders. From the plots it

follows that the interference intensities (002), (202) and (200) peaks significantly decrease, while the intensity of (100) peak rises sharply.

Structures with various types of molecular packing, especially the α, β, and γ forms or their mixtures are known to appear during polyamide crystallization. The major low temperature form of α-monocline modification has typical (002), (202) and (200) peaks, whereas (100) peak is typical for the γ-pseudohexagonal modification. Thus, the PCA mechanical activation manifests a partial polymorphous transition to the crystalline structure from the α-monocline form into a mixed structure of the coexisting α- and γ-forms (Fig. 3.13, curve 4) with typical peaks (002), (202), (200) and (100), respectively. PE softens quicker during circulation compared with PVB and PCA, and it withstands mechanical destruction better and is slower to electrify.

Thus, the results of the studies indicate that collisions of particles with the trap produce significant local deformations accompanied by the fracture of weak bonds and the appearance of free radicals. These assumptions correlate with the data of [17] proving the free-radical mechanism of the depolymerization under the effect of heat, light, ionizing radiation, and mechanical stresses.

To validate this conclusion, the typical values of viscosity has been measured. It has turned out (Table 3.11) that the particles with the size less than the size of the original particles (<71 μm) demonstrate a reduction of the characteristic viscosity upon tribodeformation. Mechanosynthesis occurs together with the mechanodestruction. This proves the stability of the process of electrification of polymeric particles during their circulation in the stream.

Table 3.11. Viscosity of polycaproamide before and after circulation during 8 hr.

Composition	Particle size, μm	Characteristic viscosity
Original	100–160	0.71
After 8 h of mechanical activation	71	0.67
After 8 h of mechanical activation	71–100	0.72
After 8 h of mechanical activation	100–160	0.90

Activation by friction alters the electrophysical characteristics of polymers, eg, the dielectric permeability of mechanically activated PP and

PVB is 10–20% higher than in the original P–(PP): ε_{init} = 2.25; ε_{act} = 2.46; for PVB ε_{init} = 3.40, ε_{act} = 4.0.

Identification of the structural and chemophysical modifications in dispersed polymers during dynamic contacts with the metal has established that the frictional activation of polymers results in the processes of partial destruction of the original crystalline structure and the chemomechanical destruction of polymeric molecules. Frictionally activated polymers manifest strong response to the reaction of thermal oxidation by the atmospheric oxygen that results from the mechanical activation, their structure becoming amorphous strongly increasing the exposure of the active centers of the polymeric macromolecules to the oxygen. Moreover, the polymer powders in the dynamic contact with metals are known to produce the electret state [7]. Because adsorption dominates in the electrically active centers on the surfaces of solids, appearance of the electret state is accompanied by a strong leap in the electrical activity leading to an intense atmospheric oxidation and humidity adsorption which, in turn, stimulates the reactivity of a polymer in tribochemical processes.

3.3. MECHANISM OF TRIBOELECTRIFICATION OF POLYMER POWDERS

Experimental results show the complexity of the triboelectrification in dispersions. Due to the cyclic nature of the process of repeating contacts along the polymer – metal boundary, the polymer powders in motion are exposed to numerous mechanical and chemophysical factors. Structural modifications of polymers, deformation of particles, alteration of their shapes, sizes and state occur due to the thermo-, chemo- and mechanodestruction, oxidation, polarization and other processes accompanied by electron, ion, dipole, and radical responses.

The particles at the moment of the dynamic contact with a metal are in a fully or partially plastic state, and as the pulverization develops they get larger. The melting of large polymeric particles at a temperature of about 473 K or any higher temperature in the entire circulation system allows us to assume the existence of high temperatures up to 10^3 K in the collisions and microcontacts. The results of the experiments have shown that high local temperatures are related to both structural transformation and electrification of polymers.

Combined radiographic structural, IR-spectrometric and thermographic studies of a wide range of dispersed polymers have revealed that the crystalline restructuring and appearance of the novel polymorphous forms are accompanied by partial introduction of the amorphous state, mechanocracking and mechanodestruction producing low molecular products due to dynamic effects. The dynamic contacts reduce the degree of crystallinity of the polymers studied by 5–10% (more strongly in the polymers with a higher rigidity of macromolecules, such as PS); also, deformation of crystallites and reduction of the interplanar distances in the crystalline lattice (by 0.1–0.2 Å for PP) are observed. A stronger reactivity of the mechanically activated polymers in the reactions of thermal oxidation by the atmospheric oxygen has been observed together with a greater optical density of the absorption bands related to the oscillation of the oxygen-containing groups in the IR-spectra of the mechanically activated polymers; the area of the exothermal peaks and their shift towards lower temperatures increase significantly (see the thermograms of the original and mechanically activated PVB and PP). The electrophysical studies also show larger values of dielectric permeability (by 10–20%) and conductivity (1.3–2 times).

Profound structural transformations of polymeric particles are accompanied by the electrification, polarization, and electretization. This explains strong electrification of the dispersed polymers circulated in the closed cycle, in as much as the velocities of 10–20 m/s make it possible to reach the potentials of 250,000 V (or 5,000 V/g of the circulating powder). As the velocity of the particles increases, the electrostatic potential grows steady, the intensity of electrification of polar polymers being much higher than that of the nonpolar polymers. Domination of the negative charge in contact with brass and its higher stability are typical also for polar dielectrics (PVB, PCA, PVC and others). The charge polarity and the electrostatic potential of a mixture of polar and nonpolar dielectrics in contact with a metal are governed by the concentrations of the components. It has been established that free spraying of powders with the particles sized 50–300 µm creates asymmetric unipolar charging, with the asymmetry becoming more pronounced as the size of particles increases; the relative quantity of negatively charged particles increases as the spraying period extends.

Both singular and multiple inversions of electrification current with the intervals up to several kiloseconds are typical for the process of static electrification during prolonged dynamic contact. They are due to both the chemophysical transformations in the polymer, specifically to the water desorption from deeper layers, and to the electron processes, such as modifications of the type of charge carriers or the direction of their travel. The structural transformations of polymer particles closely relate to the intensity of contact electrification and the emerging electret effect in a polymer. The temperature factor has been shown to be crucial in this relation. A hypothesis has been advanced to explain a stronger reactivity of polymers in tribochemical processes by the appearing electret state [13,19-21].

Based on the analysis of structural, chemophysical, and electrophysical studies and published data, it has been demonstrated that the basic postulates of the electrical theory of adhesion can be applied to the explanation of the mechanism of electrification of polymer powders during dynamic contact. The electrification of polymers is caused by the division of charges of the double electrical layer appearing at the polymer-metal boundary. When high velocity disrupts the contact a gas discharge occurs, the magnitude of the charge that remains on the particles depending on the speed, the properties of the particles material and the resistance of the environment [18–21].

The results obtained are the basis for the concept of applicability of the dispersed system to simulating contacts between block specimens. This is because the contact between two solids is always over the "spikes" with the dimensions comparable with those of surface roughness of polymer particles. Investigation of large populations of fine particles leads to higher reproducibility of experimental results. Contact processes of powder and block polymers are of a common physical nature, hence the basic characteristics typical for polymer powders can be used as effective tools for analysis of the mechanisms of triboelectrification in block specimens.

REFERENCES

1. Kraemer, H.F. and Johnstone, H.F. *Ind. Eng. Chem*, 1955, **47**, 2426–34.
2. Zimon, A.D. 'Adhesion of Dust and Powder', 2nd Edition, N.Y., 1982.
3. Loeb L. 'Static Electrification' (in Russian), Moscow–Leningrad, 1963.
4. Lushchejkin, G.A. 'Methods of Investigating Electrical Behavior of Polymers' (in Russian), Moscow, 1988.
5. Belyi, V.A. and Klimovich, A.F. *Proc. of XII Int. Conf. on Organic Coatings*, Bratislava, 1973, 8–10.
6. Goldade, V.A., Klimovich A.F. and Belyi, V.A.. *Bul. of the BSSR Acad. of Sci., Phys. & Eng.* (in Russian) , 1975, **3**, 102–107.

7. Klimovich, A.F. *Rep. BSSR Acad. of Sci.* (in Russian), 1980, **24**, no.3, 238–242, 1986, **30**, no.12, 1087–1090.

8. Frenkel, Ya.N. 'Heat Motion in Solids and Liquids and Melting Theory' (in Russian), Moscow, 1936.

9. Klementyev, N.N. 'Friction Thermodynamics' (in Russian), Voronezh, 1971.

10. Andrianova, R.L. and Pevchev, B.G. 'Strong Electrical Fields in Manufacturing Processes' (in Russian), Moscow, 1969, p. 187.

11. Lapple, C.E. and Sheperd, C.B. *Eng. Chem.*, 1940, **32**, no. 5.

12. Klimovich, A.F. in 'Improvement of Wear Resistance and Durability' (in Russian), Issue 3, Kiev, 1977, p. 47.

13. Belyi, V.A., Dovglyalo, V.A. and Yurkevich, O.R. 'Fine-Layered Polymeric Coatings' (in Russian), Minsk, 1986.

14. Bellami, L. 'IR-Spectra of Complex Molecules' (Russian translation), Moscow, 1963.

15. Piloyan, G.O. *Proc. of All-Union Symp. on Thermal Analysis Techniques* (in Russian), Moscow, 1968, 35-50.

16. Martynov, M.A. and Vylegzhanina, K.A. 'X-Ray Spectroscopy of Polymers' (in Russian), Moscow, 1972.

17. Baramboim, N.K. 'Chemomechanics of High Molecular Compunds' (in Russian), Moscow, 1971.

18. Belyi, V.A. Egorenkov, N.I., and Pleskachevskii, Yu.M. 'Adhesion Between Polymers and Metals' (in Russian), Minsk, 1971.

19. Belyi, V.A., Sviridenok, A.N., Petrokovets, M.I. and Savkin, V.G. 'Friction and Wear in Polymer-Based Materials', Oxford, 1982.

20. 'Tribology in Particulate Technology' Ed. by B.J. Briscoe and M.J. Adams. Adam Holger. Bristol–Philadelphia, 1987.

21. Dovgyalo, V.A. and Yurkevich, O.R. 'Composite Materials and Cotaing Based on Dispersed Polymers' (in Russian), Minsk, 1992.

CHAPTER 4

TRIBOELECTRIFICATION OF BLOCK AND MOLTEN POLYMER MATERIALS

Studies of the static electrification of block polymers in contact with metals allow accumulation of information on the state of surfaces layers, the processes of friction and wear in polymer-metal systems and to determine the effect of electrification on the major tribological parameters.

4.1. GENERAL CHARACTERISTICS OF TRIBOELECTRIFICATION

The extent of electrification is an effective indicator of the electrification potential, the electrical charge or the electrification current. It has been demonstrated [1,2] that the electrification potential of polymer member in metal-polymer friction pairs reaches a certain level after several seconds and then it basically does not change (Fig. 4.1, 4.2). The electrification potential of steel-polymer couples, polymetylmethacrylate (PMMA), polytetrafluorethylene (PTFE) and epoxy resins, manifest changes during charging in the initial period of time (up to 2 min). Some frictional, composite, polymer-based materials change the sign of their potential in contact with cast iron in the 10–15 minute range [3].

The changes of the potential sign are attributed to the losses of moisture by the surface layers of the polymer during the initial period of operation and to the changes of the donor-acceptor behavior. Different sequence of generation of charges has been pointed out [3 - 6] "the shaft-bearing couples have charges of opposite signs; the couples exchange the charges symmetrically; the polarity of the appearing charges changes several times." Friction on steel charges the majority of plastics

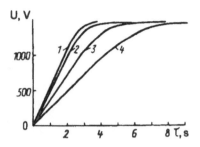

Figure 4.1. Dependence $V = f(\tau)$ for fabric-based laminate – steel friction pair at various loads, kg/cm²: 1, 3.0; 2, 1.8; 3, 1.1; 4, 0.7.

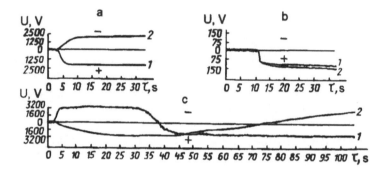

Figure 4.2. Oscillograms of total potential in friction pairs: polycaproamide – steel (a); wood particle board (with graphite) – steel (b); polymetylmethacrylate – steel (c).

negatively, so that they are conventionally called electronegative. It was demonstrated that any load modification affects the total potential. It accelerates the potential generation - as the load increases, the rate of the potential generation becomes higher until the final magnitude is reached. Greater loads increase the actual contact area. This is confirmed by the results of studies of the dependence of the electrification potential on the nominal contact area S_n at given velocity [1,2]. The tribocharge increases as the effective contact area between the two members grows [4].

The characteristics of the nature of the electrification process are the most interesting. Figure 4.3 shows variations of the potential at different velocities. The diagrams of the recorded electrification currents at low

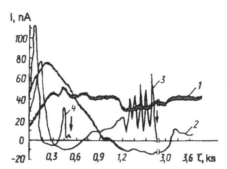

Figure 4.3. Electrification current versus operation time of polycaproamide –
metal pair at various sliding velocities v, m/s: 1, 0.5; 2, 1.0; 3, 1.5; 4, 2.0.
Pressure   is p = 0.15 MPa (arrows indicates moments of polymer rollers
disintegration).

velocity (curve *1*) indicate that initially current grows, after which it
basically remains unchanged for a long time (up to 3.6 ks).

As the friction velocity increases (curves *2-4*), the plot represents a
different pattern: 1) the electrification current direction changes; 2)
multiple "reversals" are observed; 3) the number of "reversals" increases
with the increase in friction velocity; and 4) initially the absolute
electrification current increases.

Current diagrams show the same patterns for high density polyethylene
(HDPE) – metal pairs.

Comparison of the current diagrams of polar polycaproamide (PCA)
with those of non-polar HDPE shows high positive and negative
electrification currents in the two cases in response to the conditions of the
friction contact. The polar PCA has predominantly positive polarity unlike
the non-polar HDPE. This can be explained in the following way. Strongly
hydrophilic PCA is known to demonstrate donor behavior, so that the
semiconductor is charged negatively. Therefore, dehydration may cause
positive charging, as it has already been mentioned.

Interesting data on the effect of surface roughness on the tribocharge
magnitude and sign are reported in [4]. Studies of polymethylacrylate
(PMA), quartz, solidified epoxy resin have shown the reduction of the
positive tribocharge with increased roughness. Once some critical level
roughness specific to each material is reached, the sign of the tribocharge
becomes negative (Fig. 4.4). Roughness reduces both the effective contact

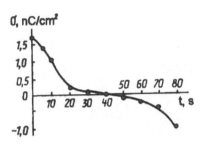

Figure 4.4. Charge σ of glass specimen at friction against cotton versus duration t of pretreatment of glass surface with sandpaper of medium granularity ($v = 10$ m/s, $N = 5$ g/cm^2)

area and the surface effects. For example, rubbing with emery paper destroys the surface layer, which produces microdefects whose number grows during friction. Because defects in solids and free radicals in polymers appear and accumulate in the process of destruction, they act as acceptors and active traps of electrons. Rough surfaces of many electropositive materials may be negatively charged by friction because the defects would trap the electrons. Therefore, surfaces with greatest concentration of electron traps have the strongest tendency to be negatively charged.

Wear-resistance of a material is strongly governed by the relaxation behavior, in addition to the durability. Hence, the life of defects produced by friction is also important, in addition to their number. The more rigid system is, the longer it takes to relax and vice versa. In friction, harder material may acquire more defects than the soft and elastic material. Friction charges the harder material negatively.

As we see, the triboelectrification conditions should determine the position of a dielectric in the triboelectrical series. First of all, it relates to the relative velocities of counterbodies and to the normal load. The results of measurements of contact velocities of different duration are reported in [5]. It is shown that the variations of the charge appear to be small compared to the stronger variations during the initial 2 second when the duration of a single contact is changed. From this it follows that the contact equilibrium requires less than 2 second to appear. Nylon-6 is charged positively, polyethylene (PE) is charged negatively. The spectral sensitivity analysis of photoemission has shown that Nylon-6 (Fig. 4.5) has the most dense states compared with PE and polypropylene (PP.) The

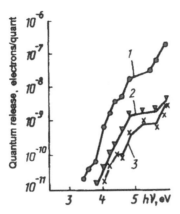

Figure 4.5. Spectral sensitivity of photoemission quantum release: 1, Nylon-6; 2, high density polyethylene; 3, polypropylene.

electron states of PE are "deeper" on average, whereas PP have "deeper" individual states.

The electron levels responsible for the charges in polymers are believed to be the states lying on the surface due to the rapid appearance of the contact equilibrium. Considering that the density of surface states of polymers is great (see Chapter 1) and it is "filled" up to energy W_i (polymer work function), the charge q (unit per area) received by the polymer through the contact with metal having the work function W_m can be determined from the expression

$$
q = \begin{cases}
e \int\limits_{W_i+\Delta V}^{W_m} N_e dE (W_m > W_i) \\
\\
-e \int\limits_{W_m}^{W_i-\Delta V} N_t dE (W_m < W_i)
\end{cases}, \qquad (4.1)
$$

where N_e is the density of filled levels; N_t is the density of traps that are assumed to be arranged over filled traps; e is the electron charge; and d is the spacing between contacting surfaces $\Delta V = dq / \varepsilon_0$.

The difference between energies $W_m - W_i$ governs the resulting direction of electron transfer, ie the charge sign.

Therefore, when a polymer is brought into contact with a metal, the electrons at the levels higher than the Fermi level in the contacting metal come to the surface, the traps with energies lower than the Fermi level are filled up with metal electrons. The resulting charge in the polymer distributes according to the following expression:

$$q = -e \int\limits_{E_v}^{W_i} N_i dE + e \int\limits_{W_i}^{E_c} N_e dE, \qquad (4.2)$$

where E_v and E_c are the "ceiling" and the "bottom" of the valence zone; W_i is the upper edge of filled surface states in equilibrium.

Lowell [6] notes that the charge density depends on the way of it is transferred from metal to polymer. The charge does not depend on the metal work function, and it is restricted to the tunneling when there is a single contact between the polymer and the metal. The charge increases to saturation depending on the metal work function in repeated or sliding contacts. During a single contact, the electrons fill up the trapping centers which are up to 50 Å, the value deep corresponding to the tunneling depth. During repeated contacts the microstructure of polymer surface is strongly deformed by metal. The trapping centers release the carriers which are redistributed between the centers, whereas new carriers favoring a stronger charge are generated .

It is also demonstrated [7] that the tendency of a material to accumulate negative electricity is maintained by defects caused by friction, rather than by those which had already existed.

A possible mechanism of tribolelectrification may be the transfer of trapping centers from place to place together with trapped electrons when deformation of polymer either does not release trapped electrons or generates new ones. When some trapping centers migrate from the surface (together with trapped electrons) into the bulk of polymer, approximately the same number of empty centers migrate from the bulk to the surface where they are filled up with electrons. As the number of repetitions of contacts increases (or sliding takes place), the centers remain within the tunneling zone continuously increasing the charge. The depth of the charge redistribution is approximately ~500, Å which is about the crystallite size in polymers.

Techniques of equivalent electrical circuits seem appealing for the analysis of triboelectrification of metal-polymer contacts. This approach was developed by Postnikov [13] to investigate dry and lubricated friction of metals.

Let's consider the following polymer–metal contact model (Fig. 4.6) [8]. In the stabilized friction mode it can be assumed that the contact occurs on the tops of "waves" on the polymer surface; the distance between separate contact spots is equal to the mean wave pitch of about 100–15,000 μm. The real metal-polymer contact area (RCA) is 3–15% [32]. Charges are generated over RCA spots with the average radius of 7–65 μm [9,10]. Frictional contacts are accompanied by the growing density of surface states, the increasing concentration of charge carriers, their accelerated migration, redistribution over the surface and in the body of the material causing the charging of noncontacting spots. Figure 4.6 *a* shows the pattern of distribution of the resulting charge over the polymer friction surface. The peaks of the charge density correspond to RCA spots. Our experimental studies and the published data [11–16] have allowed development of an equivalent electrical circuit of the polymer-metal frictional contact (Fig.4.6 b).

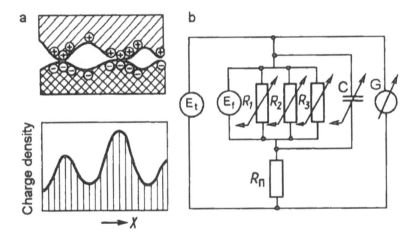

Figure 4.6. Model (a) and equivalent electrical circuit (b) of polymer - metal frictional contact.

According to this design, the metal-polymer friction pairs can be assumed to be the source of tribo- and thermal-electromotive forces (EMF) (E_f and E_t) with corresponding internal resistances. The internal resistance of the tribo-EMF is the resistance of the actual contact, which includes the resistances of areas of contacts between the polymer and the metal juvenile surface R_1, polymer – metal oxide film R_2 , and polymer – polymer film transferred on the metal R_3. The internal resistance of the thermal EMF source is the actual contact resistance connected in series with the resistance of the polymer volume R_p. The capacitor C is connected in parallel to the actual contact resistance to reflect the share added by non-contacting surface portions. In this case, the tribo-EMF is a sum of the contact component due to the difference of potentials in contacts and the component added by the mass transfer between the surfaces and sorption-desorption processes right under the surfaces. The thermal EMF is due to the temperature gradient in the contacting materials. Then, the current I_f generated in the circuit by the tribo- EMF is the algebraic sum of the following currents:

$$I_f = I_c + I_m + I_d + I_p, \tag{4.3}$$

where I_c is the contact electrification current; I_m is the current due to the motion of charged particles at frictional transfer; I_d is the current due to the sorption and desorption in the subsurface contact regions; and I_p is the reversed discharge current appearing when the friction contact is interrupted ("pulsing current").

Using the formula of Krupp, the contact charge density I_c can be determined in the case when there are three types of contact sites:

$$I_c = \frac{d}{dt}(\sigma S_R) = \frac{d}{dt} e D_S [(W_m - W_p) S_{R_1} + \\ + (W_0 - W_p) S_{R_2} + (W_{s1} - W_p) S_{R_3}]. \tag{4.4}$$

It is assumed that the density D_s of the surface states and the electron work function W_e are similar for all the sites on the polymer surface.

The analysis of the above expression indicates that in the majority of cases $[W_0 > W_p, S_{R2} > S_{R1}, (\text{at } t = 0); W_{s1} \sim W_p ; dS_{R1} / dt > 0; dS_{R3} / dt \geq 0]$, and, taking into account the effect of thermal activation of dissolution of the metal surface layer by polymer and expulsion of the metal from oxide, $[-S_{R2} / dt < 0$ (for oxygen-containing polymers it is possible that $dS_{R2} / dt > > 0)]$, ie the magnitude and the direction of current I_c is governed by the second addend during the initial period of friction and by the first addend once the system reaches the state satisfying the equality $(W_m - W_p)S_{R1} = (W_0 - W_p)S_{R2}$. Then, when $W_p > W_m$ [polyvinylchloride (PVC) – steel,

PE – steel, etc.] the direction of current I_c according to the design is negative, whereas at $W_m > W_p$ [PCA – steel, polyethylene terephtalate (PETP) – steel, etc.] it is positive. Thus, the polarity of the contact electrification current I_c changes dynamically due to variations of the contact parameters. It depends on the electrophysical characteristics (W) of contacting materials.

The magnitude of the mass transfer current I_m in the general case is determined by the algebraic sum of the currents:

$$I_m = \sum_i q_i n_i V_i, \qquad (4.5)$$

created by the particles of the i-th type during friction. Assuming that the transfer is effected by one type of charged particles, ignoring the transfer of metal particles onto the polymer, the expression for the mass transfer current can be written in the following form

$$I_m = qnV. \qquad (4.6)$$

There is an apparent correlation between the number of particles n and the wear rate. Considering that the charge created from separating polymer particles coincides with the total polymer surface integral charge, the direction of current I_m becomes opposite to the direction of I_c.

Similarly, the magnitude of I_d due to the processes of sorption and desorption can be determined. It has already been mentioned [8] that the full surface charge in the general case is a sum of charges of "slow" and "fast" states. Investigation of relaxation processes of variations of the surface charge reveals that the latter is predominantly the charge of the "fast" surface states primarily of adsorption origin. As it is reported [17], only 10^{-9} second is required for monomolecular adsorption film to appear at the room temperature and atmospheric pressure. This time is quite sufficient for origination of the adsorption film in friction and for mutual deformations of solids. Adsorption makes the density of "fast" states higher than 10^{14} cm^{-2}. Exchanges of electrons between adjacent states are possible at such concentration [18].

Experimental studies have revealed that the surface charge sign depends on the gas atmosphere composition, ie on the processes of adsorption. Based on the data in [19],one can conclude that when the quantity of adsorbed hydrogen is small, the adsorbed surface is charged positively, and with further adsorption it becomes negative. A similar pattern is produced by adsorption of water. Because the OH group acts as a donor of electrons, water adsorption increases the original positive charge or reduces the original negative charge, and eventually, the charge sign is reversed (see Section 4.2).

Hence, under certain conditions of friction, the directions of currents I_d and I_c are either opposite or coincide. The discharge current I_p results from a sharp leap in the intensity of the local electric field at the interface when the plates of the double electrical layer separate during the frictional contact breaking. Current I_p is oscillatory with a period of 10^{-6}–10^{-5} second, its direction being opposite to that of I_c.

Considering polymers as broad-zone semiconductors (the width of the forbidden zone for polyethylene is 8 eV), the expression of the thermal EMF of polymer–metal contact including the temperature T_1, the contact temperature T_2 between the polymer and the holder (polymer–metal) and the metal–holder (metal-to-metal) T_3, can be including as

$$E_t = -\int_{T_3}^{T_1}\alpha_m dT - \int_{T_1}^{T_2}\alpha_p dT. \qquad (4.7)$$

The specific thermal EMFs of the metal and the polymer satisfy the relation $\alpha_p \gg \alpha_m$ ($\alpha_p = 10^2$–10^3 μV/deg, $\alpha_m = 1$–10 μV/deg for semiconductors), hence, ignoring the contribution of the metal thermal EMF, the following expression can be obtained

$$E_t = -\int_{T_1}^{T_2}\alpha_p dT = \int_{T_2}^{T_1}\alpha_p dT. \qquad (4.8)$$

Assuming that α_p is constant within the temperature range $T_1.-.T_2$, equation (4.7) is transformed to obtain the following expression for the thermal EMF of the polymer-metal contact

$$E_t = \alpha_p (T_1-T_2). \qquad (4.9)$$

In a hot contact the thermal flux is always directed from the member with smaller α to the member with larger α. Hence, the thermal flux I_t at $\alpha_p > 0$ is directed from metal to polymer ("positive" direction).

Thus, the expression for the total electrification current, taking into account the directions of the current components, has the following form at $W_p > W_m$:

$$I_{f\text{-}} = -I_c + I_m \pm I_d + I_t + I_p, \qquad (4.10)$$

at $W_p < W_m$

$$I_{f+} = I_c - I_m \pm I_d + I_t - I_p, \qquad (4.11)$$

The analysis of expressions (4.10) and (4.11) shows that the magnitude and the polarity of the generated triboelectrification current I_f are affected both by the original behavior of contacting bodies and the combination of chemomechanical processes accompanying frictional contacts. In response to the intensity of processes of mass transfer, sorption-desorption, heat generation, and so forth, this current can have opposing directions.

4.2. CHARGE SIGN REVERSAL

The phenomenon of current reversal during friction or machining has been experimentally observed by many researchers [13]. Harper [20] explains this phenomenon for dielectrics by variations in the charging mechanism: contact charging during temperature increase transforms into friction (the terminology of Loeb [8]). The current reversal during friction of polymers has been observed under various conditions: high velocity [1,2], pressure increase [21], displacement of transferred layers from one counterface to the other [22]. The measurements of electrification current in the inverse HDPE – metal pair (Fig. 4.7) show the multiple electrification current reversals with the repetitions increasing as the velocity grows, like in PCA – metal couple (see Fig. 4.2).

The increasing friction velocity strongly reduces the time between the first and the consecutive reversals. Similarly, the electrification current curves also become modified in response to the changes in the pressure in the couple. Hence, the electrification process is affected by the pV-factor, ie by the intensity of friction in the contact.

When the load is constant, the increasing velocity increases the friction work thus heating the contacting surfaces. Table 4.1 [23] lists the temperatures of the insert in response to the friction at various velocities. Comparison of the table and the electrification current measurements (see Fig. 4.2) registered synchronously indicates that the first current reversal occurs at the temperature of the insert of 350–380 K, irrespective of the sliding velocity. The temperature of the insert (friction surface) somewhat increases and reaches the level corresponding to the electrification current reversal as the friction velocity grows. This proves that the thermal activation causes the electrification current reversal.

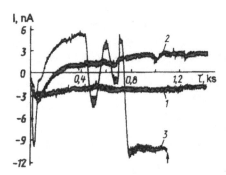

Figure 4.7. Electrification current versus time of operation for inverse high density polyethylene – metal pair at various sliding velocities v,m/s: 1, 0.3; 2, 0.8; 3, 1.5. Pressure is p = 0.1 MPa (arrows indicate moment of polymer roller disintegration)

Table 4.1. Insert temperature T, K versus time t of friction of polycaproamide – metal pairs at various sliding velocities v (p = 0.15 MPa).

t, s	v, m/s			
	0.5	1.0	1.5	2.0
30	298	299	306	320
60	303	304	312	342
100	306	308	319	368
150	310	321	330	382*
200	313	323	343	420
250	315	325	358*	452
400	316	334	367	478
600	316	338	378	–
800	317	346	385	–
1000	317	351	389	–
1100	318	356*	393	–
1200	318	361	397	–
2000	320	399	412	–
2800	320	420	431	–
3600	320	438	446	–

*Temperature corresponding to the electrification current reversal.

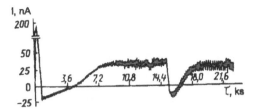

Figure 4.8.   Plot of the electrification current variations in dynamic contact between dispersed polycaproamide and metal at the velocity of 10 m/s.

A model system of dispersed polymer – metal contact was employed to investigate in greater detail a protracted friction contact between the polymer and the metal at high velocity. The analysis of the electrification current in the long time (up to 21.6 ks) friction contact between PCA and a metal indicates that the electrification current reversals repeating every several kiloseconds are typical for the model (Fig. 4.8).

The time period from the starting moment of the dynamic contact until the first current reversal and between the first and the second reversals for the dispersed PCA – metal couple are approximately equal to the corresponding intervals for the inverted PCA – metal couple at the velocity $v=1$ m/s (see Fig. 4.3, curve 2).

It is noted above that the current reversal is presumably attributed [1] to the dehydration of polymer surface layers during the initial period of friction. Our studies validate this hypothesis. It follows from the data (see Table 4.1, Fig. 4.2) that the first current reversal occurs at 350–380 K. It is known that the thermograph's analysis of PCA demonstrates water desorption at 373 K. Moreover, it also shows that for the protracted frictional contact at $v = 0.5$ m/s, although no reversal occurs, the insert is heated to at most 320 K, which is much less than the temperature of water desorption. Distler [21] proved this with his data when he noticed that adsorption of water molecules on certain active centers leads to the recharging of surface portions containing these centers. Correspondingly, the recharging also occurs during water desorption.

The analysis of the kinetic curves of electrification current for the PCA–metal pair (see Fig. 4.2) allows the qualitative and quantitative evaluation of the contribution of "fast" and "slow" surface states to the total surface charge. The equality of maximum currents at $v=1.0$ and 1.5 m/s assumes that in this case the charge is lost in "fast" states. Assuming that the charge is lost at $v=2.0$ m/s, this occurs only in 50% of the "fast" states, whereas at $v = 0.5$ m/s the charge loss partially occurs in the "slow" states. So, it can be concluded that the significant charge losses in friction can occur only from surface levels due to very fast appearance and disruption of repeated contacts. It is not correct to divide polymers into electrically positive or negative. It is shown above that the polar and the non-polar polymers in frictional contacts demonstrate both positive and negative electrification currents.

To summarize, it can be concluded that the analogs of electrification in polar, nonpolar and dispersed polymers as well as the analytical evaluation indicate that the phenomenon of current reversal may be treated as a general characteristic of electrification of polymers in friction.

Still, it should be noted that the multiple electrification current reversals require further studies, because they result from the cumulative effects of mass transfer, variations in the contact potential difference, water desorption from the surface and from the bulk of a polymer and other phenomena. The data prove the existence of the relation [8,24] between the number of defects on rubbing surfaces and their triboelectrification. The tendency of a material to accumulate negative charges is rather due to newly acquired defects during friction and wear rather than to the original defects. The charge sign reversal is governed by the wear dynamics when one or the other dielectric surface is subjected to alternating tribodestruction. In a metal-polymer contact the reversal is attributed to the conversion of debris of the frictional transfer in response to the conditions of triboelectrification [22,25,26].

4.3. TRIBOELECTRIFICATION OF POLYMER MELTS

The process of electrification of polymer melts, specifically during intense deformation, has received little attention.

Electrification of polymer composites is typical during extrusion [27]. When 1 mass % of TiO_2 is introduced into the polymer the charge becomes basically independent of the velocity, and it is smaller than during extrusion of pure PS, while going up significantly and changing its polarity when 1 mass % of ZnO is added. Introduction of 1 mass % of Al_2O_3 results in the more intricate velocity dependence: the charge reaches its maximum at smaller velocities, after which it sharply reduces with further speed reduction. This is attributed to the charge relaxation phenomenon. The effect of relaxation becomes more evident when extrusion temperature increases (Fig. 4.9), so that the polymer specific conductivity goes up.

It is also typical for polymer melts to get electrified during pneumoextrusion [28].

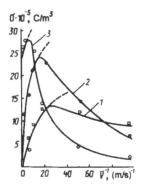

Figure 4.9. Temperature effect upon charge in extrusion of polystyrene + 1% Al$_2$O$_3$: 1, 170°C; 2,190°C; 3, 220°C.

4.4. TRIBOELECTRIFICATION OF POLYMER COMPOSITES

Because of complex structure of composites their electrification depends on the design and operation parameters of a friction pair, composition, and other factors.

The charge generated in friction contacts of polymer composites depends strongly on the nature of the counterbody metal, the contact configuration, and the filler characteristics [29].

In static contacts with metals, polymers are known [30] to have their contact charge dependent on the electron work function of the counterbody. Metals with small magnitudes of φ are highly plastic, with poor hardness, mechanical strength, they easily transfer to polymer surface with metal film becoming visible. Formation of thin metal films during friction reduces the surface resistance of polymers, resulting in a smaller generated charge. Metals with greater φ do not coat the polymer surfaces. Tables 4.2 and 4.3 show the results of studies of the effect of the contact electrical configuration on the electrification parameters, the friction force and the wear rate of direct and inverse friction pairs.

The analysis indicates that the electrification of electrically insulated polymer films is less than for those adhering to grounded metal substrates.

The grounding of the metal counterbody increases σ_{eff} for HDPE and PCA coatings.

Table 4.2. Effect of electrical circuit of contact on effective surface charge density (ESCD) magnitude of polymer in direct polymer-metal friction pairs at various sliding velocities.

Polymer	Velocity v, m/s	σ_1, nC/cm^2	σ_2, nC/cm^2
high density polyethylene	0.3	0.79	1.52
	1.4	0.92	1.39
	2	0.33	1.86
polycaproamide	0.5	0.54	1.22
	0.8	0.42	1.05
	2	0.62	0.54

Note: σ_1 is the residual ESCD of insulated polymer film generated on electrically insulated metal counterbody (roller); σ_2 – residual ESCD of polymer coating on grounded counterbody

Table 4.3. Magnitudes of effective surface charge density of polymer member electrification current I, friction coefficient μ and wear rate g at different electrical circuits of contact in high density polyethylene–steel 45 pair ($p = 0.1$ MPa, $v = 0.5$ m/s).

Friction pair	Polymer member	$\sigma_{eff} \cdot 10^5$, C/m^3	$I \cdot 10^9$, A	μ	$g \cdot 10^6$, kg/(m^{2}m)
Direct	Coating (insulated counterbody)	-3.2	–	0.19	1.45
	Coating (grounded counterbody	-4.8	1.95	0.20	1.07
	Insulated film (grounded counterbody)	-3.8	0.75	0.20	–
Inverse	Polymer shaft (insulated counterbody)	4.1	75	0.17	1.73
	Polymer shaft (grounded counterbody)	5.6	120	0.20	3.95

The grounding of the metal counterbody increases the electrification current and the friction force in direct and inverse friction pairs (Table 4.3). It affects the electrification current, the magnitude of ESCD and frictional characteristics in inverse pairs more strongly than in direct pairs. The stronger friction coefficient and stronger wear is due to the greater electrical adhesion component of the friction force, variations of local electrical fields intensity and the nature of electrical charges in the contact zone between a metal and a polymer affecting the surface dispersion of contacting bodies.

The nature, concentration, and orientation of the fibrous filler alter the electrification current by 2–15 times compared to the pure polymers. This conclusion resulted from the experimental studies of highly filled composites containing Lavsan (PETP), carbon (UV) and polyoxidiasol (TTO-3) fibers and powders of 1,3,4-polyoxidiasol and PTFE. The following compositions served as the matrices: 50 mass % of polyamide (PA) and 50 mass % of PE, PA-6 alloy with HDPE (50:50 mass %) and pure polyamide PA-6.

Initially, the polymer coatings with the fillers having different concentrations in the matrix were tested (Fig. 4.10). The analysis of the relations indicates that the more stable friction torque is typical for the mixture containing 50 mass % of PCA and 50 mass % of HDPE. Hence, this composition was selected for the matrix in further tests.

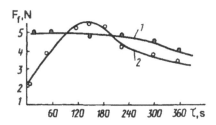

Figure 4.10. Friction force versus operating time for the following composites: 1, 50 mass % polycaproamide + 50 mass % high density polyethylene; 2, 75 mass % polycaproamide + 25 mass % high density polyethylene.

Introduction of the fibrous filer leads to the reduction of the friction force, especially in the initial period when the direction of motion and the orientation of fibers coincide, *ie* when the shaft moves along the fibers.

In this case the moment of friction stabilizes practically after 100 seconds. In case the fibers are arranged across the direction of rotation, the friction torque stabilizes after 400 second.

The anisotropy of the friction forces in the plane of orientation of the filler fibers can be explained by the fact that the shaft rotation in the direction of fibers orientation alters the actual contact area faster. The deformation components of the friction force during the sliding of the steel indenter against the oriented PTFE are known [31] to be independent of the sliding direction, whereas the adhesion component in friction across the molecular chains is almost 20% greater than the one along them. The relation between the adhesion component and the sliding direction is attributed to the stronger shear resistance (by 45%) when specimens are tested in transverse orientation.

A comparative analysis indicates that the composites reinforced with the oriented fibers have the friction force initially 30–40% less than non-reinforced ones, *ie* the results correlate with those shown above. During initial displacement (at the starting moment) the contact most probably runs over the matrix material, and after some period of friction the fibers become exposed. This process is faster when the fibers are arranged longitudinally rather than transversely. Investigation of friction of oriented PTFE has shown [31] about 30% higher friction coefficient in the sliding across the chains than along them. Similar results are reported in [32,33].

Thus, introduction of oriented fibers into the composite is, to some extent, leads to the effects similar to the ones for the oriented polymer, due to the reduction of the actual contact area, deformation and adhesion components of the force of friction. Friction current characteristics of such compositions demonstrate a different pattern (Fig. 4.11). When fibers are arranged across the sliding direction, the triboelectrification current acquires the "peaked" dependence, *ie* it reverses and produces two optima when fibers run along. The extreme zones and the inverse transition basically coincide, being within 300–400 second of each other. The contact zone temperature reaches 350–380 K.

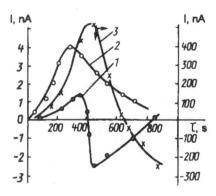

Figure 4.11. Kinetics of electrification of composites with fibers (Lavsan) (1,2) and dispersed (1,2,3–polyoxidiasol powders) (3) fillers: matrix 50 mass % polycaproamide and 50 mass % high density polyethylene. Filler orientation longitudinal (1); transverse (2).

Manifestation of the inverse transition of the specimens with the longitudinal arrangement of fibers proves that the   water desorption proceeds faster compared to the   transverse fibers. Earlier studies [23] have demonstrated that   the dehydration of polymer layers, causing the electrification current reversal, occurs exactly at 350–380 K.

Figure 4.11 shows that introduction of the   dispersed filler (curve 3) produces a more intense electrification (the triboelectrification current increases almost   by two orders of magnitude) compared to the   fibrous fillers (curves 1,2).

The data on the effect of the nature, concentration, and the orientation of the   fibrous fillers electrification rate and the coefficient of friction are given in Table 4.4. It shows that the highest electrification current is produced by the carbon fibers. The   higher electrification currents are produced by   the transverse arrangement of fibers. When the filler concentration is increased   from 5 to 20% the electrification current reduces for TTO-3 and PETP fibers, while increasing by about 10% for carbon fibers and   stabilizing afterwards.

Table 4.4. Electrification characteristics of polymers and composites with fibrous fillers.

Composition		Specimen	Friction	Electrification
matrix	filler (fiber orientation)	thickness, mm	coefficient	current *I*, nA
Polyamide (PA)	–	380	0.239	33
PA–90% Penton–10%	Lavsan (longitudinal) Lavsan (transverse)	260 320	0.215 0.200	30 57
PA–90% Penton–10%	Carbon (longitudinal) Carbon (transverse)	570 583	0.222 0.243	176 183
PA–90% Penton–10%	TTO-3 (longitudinal) TTO-3 (transverse)	408 590	0.177 0.154	116 53

The results of the studies of the effect of the electrophysical nature and the concentration of the dispersed filler in PE coatings correlate to the magnitude and kinetics of tribocharge relaxation in friction with metals. It was found that HDPE coatings and composite coatings based on PE are charged negatively in contact with metals. Introduction of graphite leads to the monotonous reduction of on effective surface charge density (ESCD.) The relationship between ESCD and the concentration of nickel and PTFE has a more complex pattern. The appearance of the peak in ESCD dependence is explained by the effect of the surface separating the filler from the polymer and by greater number of defects of the polymer material when it contains the filler. This conclusion correlates with the earlier analyzed effect of the greater density of surface states and the appearance of Maxwell-Wagner polarization due to the larger area of the separating surface when a filler is introduced.

Thus, the tribophysical phenomena are mainly governed by the nature, concentration, and orientation of the composite components. This should be taken into account when developing and employing tribosystems with composites. Their behavior is governed primarily by the combination of the electrophysical and the frictional behavior of contacting materials, friction conditions and reversals of the electrification current and the tribocharge. In the general case, the rate of electrification in the friction contact between a polymer and a dielectric is less than that one in the contact between a polymer and a metal. The system of the nonpolar HDPE and the polar PCA is closer to the polymer–metal one in its nature and characteristics of electrification. The processes of mass transfer and frictional heating of contacting materials play the significant part in tribocharging. The mechanisms of contact and frictional electrification due to strong charging because of frictional mass transfer dominate in the cases of severe wear of contacting materials.

4.5. EFFECT OF ELECTRIFICATION ON FRICTION

During the operation of metal-polymer friction pairs the intensive electrophysical processes evolve, ig accumulation of injected charge carriers and electric polarization of the polymer during electrification.

The electrostatic component of adhesion in a frictional contact of polymer and metal results from triboelectrification, which can significantly increase the friction force. This is explained by the effect of the charge with the ESCD σ_{eff} on the polymer surface. In the case of the parallel-plate capacitor it increases the nominal pressure p by $p_a = \sigma_{eff}^2 / (2\varepsilon_0)$. The values of p_a estimated for various σ_{eff} are given in Table 4.5.

The analysis of the data in the table indicates that the mutual attraction of surfaces may increase the applied pressure in response to the nominal pressure and ESCD. It has been experimentally established that the reduction of the residual tribocharge σ_{eff} by 1.5 times (reduction of the electrification current) by electrical insulation reduces the friction coefficient by 15%, the wear rate of the polymer and metal members in the inverse couple, such as PE – steel 45, by 1.3 and 2 times, respectively.

Table 4.5. Variations of nominal pressure in response to effective surface charge density, σ_{ef} .

σ_{ef} , C/m^3	p_s,Pa	p_s/p , %		
		$p = 0.05$ MPa	$p = 0.1$ MPa	$p = 0.5$ Mpa
$1\cdot10^{-6}$	5.6	0.01	0.005	0.001
$5\cdot10^{-6}$	141.2	0.28	0.141	0.028
$1\cdot10^{-4}$	564.9	1.13	0.564	0.113
$5\cdot10^{-4}$	14100	28.2	14.1	2.82
$1\cdot10^{-3}$	56500	113	56.5	11.3

The relation between the electrification rate and the friction parameters is evidenced by the studies of PCA electrification. Figure 4.12 shows the simaltenous measurements records of the electrification current and the friction force in direct PCA – metal pairs showing a definite correlation between the I–τ and F–τ dependencies. Initially, these two dependencies have a smooth pattern, and after current reversal (1.3–1.4 ks) they acquire a synchronous leaping pattern. As the contact extends in time, the friction force increases. Then the I–τ dependence passes through the maximum which is close to the glass transition temperature of PCA $T = 323$ K.

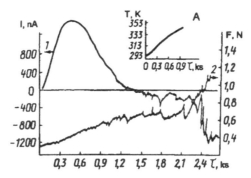

Figure 4.12. Plots of synchronous measurements of kinetics of electrification current (1) and friction force (2) in direct polycaproamide – steel pair (p = 0.1 MPa, v = 2.0 m/s); A, kinetics of variations of metal insert temperature at friction.

Studies of the relation $\sigma_{eff} - \tau$ measured after 30–40 s of friction have revealed that the maximum of σ_{eff} extrapolated to the time of operation is below the glass transition temperature of PCA (Fig. 4.13 a, curves $1,2$).

It has been demonstrated [34] that σ_{eff} has the maximum depending on the friction velocity, but remaining close to   the glass transition temperature of the polymer (Nylon-6). Electrification is assumed to be dependent on the molecular motion in polymers.

Similarity of $F-\tau$ and $I-\tau$ dependencies should ideally manifest itself with longer time of friction. Yet, due to the effect of the temperature in friction and accompanying chemophysical processes, the plot of $I-\tau$ shows the peak , and then the curve goes down and the reversal occurs.

Comparison of the ESCD temperature dependence $\sigma_{eff} - T$ obtained at the   friction of   the preheated PCA coatings with the temperature dependence of the specific bulk electric resistance of the coating $\rho_b - T$ (Table 4.6) concludes that the relation $I-\tau$ is mainly determined by the structure and the properties of a polymer, specifically the temperature dependence of the polymer electric conductivity.

Table 4.6.   Magnitudes of bulk resistance of high density polyethylene and polycaproamide coatings at various temperatures.

ρ_b, Ohm m	T, K						
	293	323	343	363	383	403	423
high density polyethylene	$3.6 \cdot 10^{14}$	$3 \cdot 10^{14}$	$2.8 \cdot 10^{14}$	$7.5 \cdot 10^{13}$	$7 \cdot 10^{13}$	$3.7 \cdot 10^{12}$	–
polycaproamide	$2.9 \cdot 10^{13}$	$2 \cdot 10^{13}$	$1.6 \cdot 10^{13}$	$2.2 \cdot 10^{12}$	$2.1 \cdot 10^{11}$	$8.7 \cdot 10^{9}$	$6.5 \cdot 10^{8}$

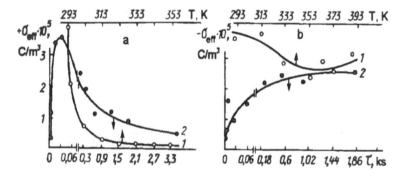

Figure 4.13. (a) Effective surface charge density of polycaproamide and (b) high density polyethylene (1) coatings versus preset temperature of polymer surface at friction and (2) duration of friction against steel (v = 0.5 m/s, τ = 0.18 ks (1); p = 0.18 MPa (a), 0.1 MPa (b)).

This is proven by the experimental relation σ_{eff} –τ and σ_{eff} –T during friction of PE coatings (Fig. 4.13 b), as well as by the temperature dependencies of ρ_b –T of PE coatings (Table 4.6). The table shows that, when T reaches 403 K or when it is close to the melting temperature of a polymer, the electric conductivity of HDPE increases by the two orders of magnitude. It initially results in some reduction of the maximum of σ_{eff} with subsequent restoration of its value (Fig. 4.13 b).

Again, as T reaches 403 K the electric conductivity of PCA increases by the four orders of magnitude. Apparently, this results in a higher density of surface states, filling them up with charge carriers and increasing the electrification rate. Yet, at the same time, the electrical conductivity and molecular mobility in the polymer increase, resulting in the smaller magnitude of σ_{eff}. Therefore, the maximum the σ_{eff} –τ and I–τ curves reduces insignificantly for HDPE and more sharply for PCA which is explained by the behavior of the latter. Analyzing the I–τ and F–τ relation and remembering the presence of the maximum on the current curve near   the glass transition temperature, a strong structural responsiveness of the electrophysical behavior in friction should be noted.

REFERENCES

1. Bilik Sh.M. 'Metal-Plastic Friction Pairs in Machinery' (in Russian), Moscow, 1965.
2. Tsurkan, V.P. in 'Plastics in Sliding Bearings' (in Russian), Moscow, 1965, p. 75.
3. Georgievskii, G.A., Lebedev, L.A. and Borozdinskii, E.M. in 'Electrical Phenomena at Friction, Cutting and Lubrication of Solids' (in Russian), Moscow, 1973, p. 12.
4. Vladykina, T.N. and Toporov, Yu.P. in 'Non-Equilibrium Processes in Dielectric Materials' (in Russian), Moscow, 1983, p.180.
5. Murata, Yu. *Japanese Journal of Appl. Phys.*, 1979, 18, no.1, 1–8.
6. Lowell, J. and Rose-Innes, A.A. *Advances in Physics*, 1980, 29, no.6, 947–1023.
7. Vladykina, T.N., Toporov, Yu.P. and Luchnikov, A.P. *Soviet Journal of Friction and Wear*, Allerton Press, N.Y. , 1988, 9, no.3, 117–121.
8. Klimovich, A.F. and Mironov, V.S. *Ibidem*, 1985, 6, no.5, 796–806; no.6, 1026–1033.
9. Kragelskii, I.V. and Alisin, V.V. 'Friction, Wear and Lubrication. Reference Manual' (in Russian), Moscow, 1978.
10. 'Fundamentals of Tribology'. Ed. by A.V. Chichinadze (in Russian), Moscow, 1995.
11. Gode, M. *Soviet Journal of Friction and Wear*, Allerton Press, N.Y., 1991, 13, no.1, 27–42.
12. Drozdov, Yu.N., Pavlov, V.G. and Puchkov, V.N. 'Friction and Wear under Extreme Conditions' (in Russian), Moscow, 1986.
13. Postnikov, S.N. 'Electrical Phenomena in Friction and Cutting' (in Russian), Gorkiy, 1975.

14. Adams, M.J., Briscoe, B.J. and Pope, L. in 'Tribology in Particulate Technology', Adam Hilger. Bristol–Philadelphia, 1987., p. 8.
15. Balachandran, W. in 'Tribology in Particulate Technology', Adam Hilger. Bristol–Philadelphia, 1987., p. 135.
16. Bahadur, S. and Tabor D. *Wear*, 1984, 98, 1–13.
17. Bikerman, J.. *Wear*, 1976, 39, no.1, 1–13.
18. Kiselev, V.F. and Krylov, O.V. 'Electron Phenomena in Adsorption and Catalysis on Semiconductors and Dielectrics' (in Russian), Moscow, 1979.
19. Volkenshtein, F.F. 'Chemophysics of Semiconductors Surfaces' (in Russian), Moscow, 1973.
20. Harper, W.R. 'Contact and Frictional Electrification', Oxford, 1967.
21. Distler, G.N. and Moskvin, V.V.. *Rep. USSR Acad. Sci.* (in Russian), 1971, 201, no.4, 891–893.
22. Vladykina, T.N., Toporov, Yu.P. and Luchnikov, A.P. *Soviet Journal of Friction and Wear*, Allerton Press, N.Y., 1988, 9, no.3, 117–121.
23. Klimovich, A.F. and Mironov, V.S. *Soviet Journal of Friction and Wear*, Allerton Press, N.Y., 1981, 2, no.3, 128–131; no.4, 113–118.
24. Vladykina, T.N., Deryagin, B.V. and Toporov, Yu.P. *Surface. Physics, Chemistry, Mechanics* (in Russian), 1989, 9, 149–151.
25. Evdokimov, Yu.A. and Kolesnikov, V.I. *Journal of Friction and Wear*, Allerton Press, N.Y., 1993, 14, no.2, 127–133.
26. Sviridenok, A.I. In 'Tribology in the USA and the Former Soviet Union:Studies and Applications'. Ed. by V. Belyi, K.C. Ludema, N.K. Myshkin, Allerton Press, N.Y., 1993, p. 157.
27. Taylor, D.M. and Lewis, T.J. 'Electrification of Polymers during Extrusion', Univ.College of North Wales Bemdor, Caernarvonshire, 1982.
28. Shustov, V.P., Sviridenok, A.I., Sukanevich, A.V. and Gajduk, V.F. *Proc. of Saving Resources and Ecologically Clean Technologies Conf.* (in Russian), P. 2, Grodno, 1995, 217–222.
29. Mironov, V.S., Malozemova, T.I. and Klimovich, A.F. *Proc. of Friction and Wear of Composite Materials Conf* (in Russian), Gomel, 1982, 56–57.
30. Lee, H.L. in 'Polymer Wear and It's Control', S. 287, Washington, 1985, p. 27.
31. Bowden, F. and Tabor, D. 'The Friction and Lubrication of Solids', Clarendon Press, Oxford, 1964.
32. Belyi, V.A., Sviridenok, A.N., Petrokovets, M.I. and Savkin, V.G. 'Friction and Wear in Polymer-Based Materials', Oxford, 1979.[33]
33. Vinogradov, G.V. and Bartenev, G.M. *Rep. of the USSR Academy of Science*, 1968, 180, 1082.
34. Ohara, K.. *Journal of Electrostatics*, 1978, 4, 233–246.

CHAPTER 5

TRIBOELECRET STATE OF POLYMERS

Intense electrophysical processes in dielectric materials employed in metal-polymer friction pairs generate their triboelectret state. It is strongly governed by the electrostatic interactions between contacting surfaces and the magnitude of the adhesion component of the force of friction.

5.1. GENERATION OF TRIBOELECTRETS AND THEIR EFFECT ON FRICTION AND WEAR

The analysis of the energy balance in friction indicates that a portion of the mechanical energy goes to generation of the triboelectret state in a polymer during the run-in and to maintaining it in the stationary conditions of friction. Indications of the appearance of the charge of 1–10 $\mu C/m^2$ in specimens subjected to friction, as well as persistence of the charge up to 160 days and the presence of a spectrum of thermally stimulated currents (TSC) have allowed identification of the electret state of polymers during friction in fluids, such as dielectrics and electrolytes [1-3]. It has been noted that the friction in fluids with various electric conductivity increases the general effect of adsorption processes upon the magnitude of the generated charge more than the electric resistance of the medium. The appearing electret state (ES), in its turn, affects the chemophysical processes in friction of polymers. In particular, the adsorption processes evolve primarily on the surface of electrically active centers of solids, and, therefore, the appearing ES makes polymers more reactive in tribochemical processes [2].

Appearance of the ES in the traditional polymer electrets may be attributed to the accumulation of charges due to the injection and the polarization caused by orientation of dipoles. The verification of this assumption requires the study of the electret effect in nonpolar high density polyethylene (HDPE) and polar polyvinylchloride (PVC) polymers. In order to monitor the effect of temperature upon the ES kinetics, the friction test must be discontinued after a certain fixed temperature is reached in the friction zone.

The analysis of TSC-graphs of HDPE coatings (Fig. 5.1) shows that two trapping levels exist with corresponding low temperature peak within the range of T = 348–358 K and high temperature peak within the range T = 388–403 K. The depth of traps (the activation energy) estimated for the first level using the "initial rise" technique [3] turned out to be within the range of 1.07–1.8 eV (10.33–173.7 kJ/mol) in response to the temperature T in the friction zone. Trapping centers in polyethylene (PE) are known to appear when the activation energy is within the estimated range causing structural modifications.

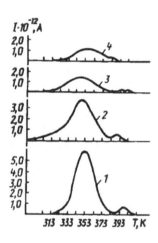

Figure 5.1. Thermally stimulated currents diagrams of high density polyethylene coatings after friction against steel counterbody (p = 0.2 MPa, v = 1.0 m/s) at various temperatures in friction zone: 1, 323 K; 2, 333; 3, 353; 4, 373 K.

Thus, considering that PE is nonpolar, the TSC values are positive and the position of the peaks on the temperature scale correspond to the relaxation transitions and estimated activation energy levels, so it can be asserted that the generation of ES in PE is due to the injection processes. The observed TSC spectra in PE are caused by the charge liberation when separate links, segments, and larger kinetic units are displaced [4].

Unlike the nonpolar partially crystalline PE, a negative polarity peak near the glass transition temperature T_g in the TSC spectrum of the amorphous polar PVC (Fig.5.2) proves that in this case the ES is generated by the injection of charge carriers from the metallic counterbody assisted by dipole polarization. The negative peak (T=343–348 K) within the glass transition range (T_g of PVC is 345 K) is due to the molecular mobility, the amorphous nature of the phase and the dipole segment α-process. The positive peak observed above T_g, *i.e.* at 358–363 K, is caused by the relaxation of space charges injected during friction due to the conductivity.

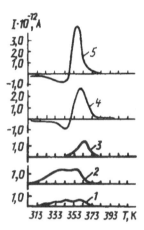

Figure 5.2. Thermally stimulated currents diagrams of polyvinylchloride coatings after friction (p = 0.2 MPa, v = 1.0 m/s) at various temperatures in friction zone: 1, 303 ; 2, 313; 3, 323; 4,- 333; 5, 343 K.

Each relaxation process has its typical duration that depends on the temperature according to the well-known equation of Boltzmann-Arenius:

$$\tau_i = B_i \ \exp(V_i \ / \ kT) \tag{5.1}$$

The relationship between the effective volume of the kinetic unit W_k and factor B is expressed by the formula

$$W_k = [B^6 \ (6kT/\rho)^3]^{1/5} \tag{5.2}$$

Based on the TSC-diagrams the parameters of relaxation (according to Fig. 5.2) for the negative peak have been determined using equations (5.1) and (5.2). Table 5.1 shows the results.

Table 5.1. Parameters of relaxation in polyvinylchloride coatings subjected to friction.

Temperature in contact zone T_c, K	Activation energy U_i, kJ/mol	Relaxation time τ at T=300 k, s	Dimensional factor V, s	Kinetic unit volume W_k, m^3
333	69.9	13217	$7.2 \cdot 10^{-8}$	$4.1 \cdot 10^{-23}$
343	59.8	9013	$2.8 \cdot 10^{-7}$	$4.3 \cdot 10^{-22}$
353	63.7	8715	$7 \cdot 10^{-8}$	$6.4 \cdot 10^{-23}$

* Activation energy is determined by the "initial rise technique".

This analysis indicates that the values of the estimated parameters correspond to the λ-process of relaxation. The relaxation time according to [5] is $\tau_i = 10^2 - 10^4$ s at T=300 K, the activation time is V_i=30–50 kJ/mol, the kinetic unit volume is W_i=$10^{-23} - 10^{-24}$ M^3. Also, according to the negative peak position corresponding to the glass transition temperature (Fig.5.2, curve 4), the absence of the negative peak in the experiments, once they were suspended before this temperature is reached, (Fig.5.2, curves 1–3), proves that the α-process of relaxation takes place in this case. Apparently, the observed abnormality is caused by the specific friction effects when both macro- and microvolumes of contacting bodies get involved, with their sizes after friction transfer reaching $10^{-19}...10^{-20}$ m^3. It is accompanied by the orientation and elongation of molecular chains in the direction of friction. Hence, the negative peak on the TSC-graphs of PVC specimens is due to the α-process of relaxation in friction with larger values of parameters usually because of the dispersion and the orientation by friction.

Thus, the ES in PVC coatings in a direct couple friction on the steel counterbody appears in the following way: when members come into contact and go apart, the charge carriers are injected into the polymer and localize over the trapping centers or surface states. As the temperature of the contacting bodies increases, the molecules are liberated and dipole groups get oriented in the direction of friction and the injected charge field. The friction process is known to be accompanied by the orientation of segments of the molecules of the polar groups in the main and the transferred material in the direction of the force of friction which may cause polarization. It is caused by the mechanical (the effect of the normal and tangential forces) and electrostatic (the inner field effect produced by electrization) factors. The degree of this effect upon polarization depends on the structures and physical behavior of contacting materials [3-9]. The TSC-graphs (Fig. 5.2) allow to conclude that the parameters of the negative (dipole) peak governed by the numbers of oriented dipole groups, are determined primarily by the space charge magnitude injected into polymer by friction. It means that the orientation of dipole groups in PVC is produced by the internal field created by injected charge carriers (the electrostatic factor) and the orientation produced by friction.

These characteristics of ES generation in polymers have a quite general nature. But the kinetics of the ES evolution and the electrification kinetics should be given special consideration.

It has been demonstrated earlier that the current inversion, as a general characteristic of polymers triboelectrification, proves the intricacy of these processes and their relation to the chemophysical processes in polymer surface layers. The TSC-graphs for PVC coatings during various stages of the electrification (Fig. 5.3) indicate that the moments of time before and after inversion coincide with the earlier established temperature effects in the friction zone upon the ES generation in friction (Fig. 5.2, curves $1,2$ and $4,5$), because the temperature in the contact zone (Fig. 5.3, curve 1) T_h=306 K corresponds to that one in experiments 1 and 2 (Fig. 5.2, curves $1,2$); T_h=346 K (Fig. 5.3, curve 3) correspond to the temperature in experiment 5 (Fig. 5.2, curve 5).

A typical feature of the TSC spectrum at the moment of inversion is a significant reduction (by 5–10 times) of the intensity of the positive peak at 353–363 K. The notion of "the moment of inversion" is conventional, but only the moment of approaching to such inversion can be properly referred to in experiments. Proceeding from the temperature in the contact zone, T_k=328 K (Fig. 5.3, curve 2) is an intermediate temperature for experiments 3 and 4 (Fig. 5.2), and when T_k magnitudes are equal to 323 and 333 K, respectively, the pattern of TSC-graph 2 (Fig. 5.3) can be considered intermediate between graphs 3 and 4 (Fig. 5.2). Yet, a significant reduction of the positive peak intensity at 353–363 K proves that both the surface layers (e.g., desorption) and the bulk of the polymer experience strong transformations [10]. Hence, the tribological behavior closely relates to the appearance and relaxation of ES in a polymer.

The positive peak at 343 K (Fig. 5.2, 5.3, curves $1,2$) should be noted separately. The analysis indicates that it occurs in the case when the friction zone temperature is much less than the glass transition temperature. Apparently this peak is due to the depolarization under the effect of the field of injected charges, and when the thermally stimulated discharge appears, it has nothing to do with structural modifications of the polymer during electrification and electretization.

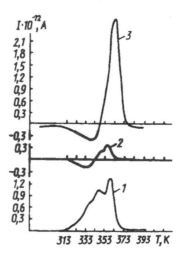

Figure 5.3. Thermally stimulated currents diagrams of polyvinylchloride coatings for different stages of triboelectrification process: 1, prior to electrification current inversion; 2, at the moment of inversion; 3, after inversion.

Triboelectretization in the contact between two dielectrics (polymer-polymer) deserves some additional discussion. HDPE was tested in the form of coatings 350–400 µm thick made of the PE powder melt by pressing it at 5 MPa during 0.3 ks at 473 K against the aluminum foil substrates. The dielectric counterbody was made from polycaproamide (PCA), HDPE and polytetrafluorethylene (PTFE). The original roughness of rollers (0.7–0.45 µm) was prepared with abrasive paper. The tests were calibrated as a shaft on a partial insert at a nominal pressure 0.1 MPa, the friction torque, the temperature of the specimen and current parameters were registered synchronously. The wear rate of coatings and rollers were evaluated by weighing after continuous operation during 1.8 ks. The air relative humidity varied within 65–75%. The effective surface charge density (ESCD) characterizing the degree of triboelectrification was measured using the compensation technique with the help of a vibrating electrode. TSC of the coatings was registered using aluminum electrodes at a linear heating rate 2.5 deg/min. The results were averaged relative to the air humidity variations.

The results have established (Table 5.2) that the magnitude and the polarity of the residual, and, hence, the generated tribocharge on the coatings strongly depend upon the dielectric behavior of the counterbody material and the friction mode.

Table 5.2. Effective surface charge density (σ, $\mu C/m^2$) of high density polyethylene coatings triboelectrified at various friction velocities in couple with dielectric counterbody (relative humidity 65–75%).

v, m/s	Counterbody material					
	Polycaproamide		High density polyethylene		Polytetrafluorethylene	
	n_{1-}/n_{1+}	σ_{1-}/σ_{1+}	n_{2-}/n_{2+}	σ_{2-}/σ_{2+}	n_{3-}/n_{3+}	σ_{3-}/σ_{3+}
0.3	100/0	57/–	36/64	6/11	71.29	20/5
0.5	100/0	47/–	22/78	10/17	67/33	28/11
0.8	100/0	60/–	71/29	11/5	50/50	28/13
1.1	100/0	46/–	86/14	14/2	80/20	31/19
1.4	100/0	58/–	100/0	8/–	0/100	–/9

Note. All polycaproamide specimens were positively charged.

The HDPE coatings in contact with the polar PCA at the friction velocities used become negatively charged. The same coatings in contact with the nonpolar PTFE and HDPE acquire an unstable residual tribocharge: the specimens may become both positively and negatively charged under the same friction conditions. Table 5.2 demonstrates that the growing friction velocities change the ratio between the negatively and positively charged specimens and between the mean ESCD magnitudes. The dominating charge polarity changes from positive to negative when the counterbody is HDPE and vice versa when it is PTFE. The ESCD inverts at friction velocities 0.5–0.8 and 1.1–1.4 m/s, respectively. The mean ESCD magnitudes change most extremely in contact with PCA, HDPE and PTFE as the friction velocities grow (passing through the maximum) satisfying the relations

$$\sigma_{1-} > \sigma_{3-} > \sigma_{2-}; \quad \sigma_{2+} \approx \sigma_{3+}.$$

The above friction couples differ by the friction coefficients, the temperature of specimens and the wear rates (Table 5.3). The above parameters are typically minimal for HDPE – PCA systems and the

TRIBOELECRET STATE OF POLYMERS

maximal for HDPE – PTFE systems. Tables 5.2 and 5.3 show that the higher wear rates and the higher friction velocities accelerate the charge inversion.

Table 5.3. Tribological characteristics of the high density polyethylene – dielectric friction pairs (p = 0.1 MPa, v = 0.8 m/s, t = 1.8 ks).

Counterbody material	T, K	Friction coefficient	Wear rate I_g, kg/(m^2 m)	
			high density polyethylene coating	polymer roller-counterbody
Polycaproamide	305	0.15	$1.72 \cdot 10^{-7}$	$1.65 \cdot 10^{-7}$
High density polyethylene	333	0.21	$6.92 \cdot 10^{-7}$	$4.20 \cdot 10^{-8}$
Polytetrafluore-thylene	346	0.24	$5.50 \cdot 10^{-6}$	$3.72 \cdot 10^{-6}$

Because these polymers have different work functions, different surface (water adsorption) and frictional behavior, it can be assumed that the charge inversion is due to the change of the dominating charging mechanism when the friction velocity is altered.

It has been demonstrated above that the polymer electrets are a classic example of disordered systems. Also, the friction of polymers can be assumed to be an integral process accompanied by a wide variety of electrophysical phenomena (see Chapter 1). Apparently, they interact with a possible feedback, self-control and self-restoration [11,12].

Possibilities of the appearance of an autoregulating system in the process of evolution and relaxation of the ES of polymers in friction is a worthwhile study. Therefore, the cascading (intermittent) friction mode was used for experimentation with the electrophysical parameters. The experiment was set up as follows: HDPE films 300–320 µm produced by hot pressing were subjected to the cyclic friction at $p = 0.1$ MPa and v = 1 m/s during variable time periods t in the cascading mode with the interval τ between cycles. Common techniques were applied to check the parameters of the ES of coatings after friction by measuring ESCD and TSC.

Initially, the effect of the friction duration on the triboelectrification was studied. The specimens were subjected to friction during one cycle lasting 60, 300, 600, and $1.2 \cdot 10^9$ s. The TSC spectra of the coatings manifested two typical peaks: the low temperature one in the range of 303 K and the high temperature one in the range 383–403 K, the high temperature peak tending to bifurcate as the friction extends.

A different pattern is observed when the number of cycles is varied together with the intervals between them. The increasing number of cycles reduces the high temperature peak, *i.e.* the magnitude of the electret charge as the variations of the ESCD evidence this under different friction conditions. As the interval is increased, the ESCD grows noticeably. This can be explained by the fact that the continuous and the concurrent electrification processes produce the electret charge with consecutive appearance of the space charge domain (SCD), and because the existing electron states in the SCD field are ionized, the emission phenomena occur and the charges are relaxed. Such relaxation primarily occurs in the interval between the cycles: the charges in the fast surface states are the first to reduce followed by those in the slow states. The charge relaxation becomes faster as the interval grows. Subsequent friction cycles cause the filling of the vacant high-energy traps and the ESCD goes up.

So, it can be concluded that such unsteady systems are capable undergoing self-organization when collective, cooperative effects play a substantial role in the polymer electrets.

A sample of 100 μm HDPE coatings produced by hot pressing against aluminum foil substrates were thermally electrified using the dielectric PTFE spacers. The technique described in [10] was applied to investigate their physical and mechanical behavior in friction.

It has been established that the electret state generated in advance strongly affects the wear of PE coatings (Fig. 5.4). The mass wear rate depends on the ESCD of both signs, which have nearly parabolic shape.

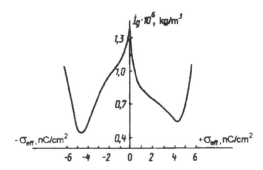

Figure 5.4. Mass wear rate I_g vs. effective surface charge density magnitude and the sign of thermoelectretized polyethylene coatings (p=0.5 MPa; v=0.5 m/s).

The radiographic and infra-red (IR) spectroscopy evidence some increase (by 2.5%) of the crystallinity of HDPE coatings after thermal electretization; no other significant chemophysical or structural modifications have been detected. Treatment of PE in a constant electrical field $3 \cdot 10^7$ V/m is reported to increase the crystallinity by 2% [13], correspondingly making the crystallites smaller by 12%. Considering these facts and the earlier detected reduction of the friction coefficients, and the fact that the improved wear resistance of polymers when spherolites structures become smaller has been reported [14], electretization can be assumed to be one significant parameter affecting friction, wear, and transformations of supermolecular structures.

Analyzing Fig. 5.4, it should be pointed out that the frictional behavior is impaired when the ESCD becomes extreme reaching the value of $(4...5) \cdot 10^{-9}$ C/cm^2 which may be attributed to the negative effect of the electret "excessive" charge. Decoration patterns of the charge distribution over the surfaces of PE thermoelectrets before and after friction prove that fact (Fig. 5.5).

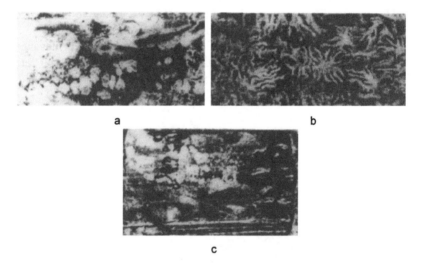

a   b

c

Figure 5.5. Charge distribution over surfaces of thermoelectretized polyethylene coatings: a, original specimen, $\sigma_{eff} = +3.64 \cdot 10^{-9}$ C/cm^2; b, original, $\sigma_{eff} = -8.72 \cdot 10^{-9}$ C/cm^2; c, after friction, $\sigma_{eff} = +0.52 \cdot 10^{-9}$ C/cm^2 (original value $\sigma_{eff} = 3.64 \cdot 10^{-9}$ C/cm^2). Dark background is a region of negative charge distribution.

The electrophotographic powder was applied to decorate surfaces and visualize charge distribution patterns. The powder was charged positively in the experiments. The powder was deposited on negatively charged surfaces to make their electrically active components visible.

The figure 5.5 shows that originally the clusters of decorating particles are diffused manifesting a peculiar "dendrite" pattern typical for discharge processes (see Fig.5.4 *b*) of the ESCD of a specimen with a specific dendrite pattern reaching $8.72 \ 10^{-9}$ C/cm^2, *i.e.* almost 2 times higher than the extreme ESCD in Fig. 5.4, corresponding to the region of a strong wear leap. Apparently, the active discharge processes on the polymer electret surfaces have a negative on its frictional behavior. These decoration patterns evidence a mosaic distribution of positive and negative charges both before and after friction, though this mosaic pattern is much weaker after friction (Fig. 5.5 *c*): there is a typical degree of orientation in the direction of friction and a homogenous charge distribution.

Thus, the decoration patterns prove, first of all, that the friction is a characteristic process of the texturing of surfaces which favors the ordering of the charge distribution.

Studies have confirmed that the electret state of PE coatings produced in advance is strongly transformed by the contact between polymers and metals under the effect of local high intensity electrical fields appearing in the microcontact, due to the injection of charge carriers during contact electrification and heat liberation in the friction zone. Figure 5.6 shows how the residual ESCD depends upon the friction path and how it converts

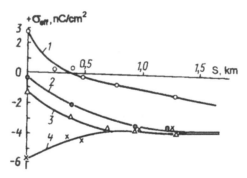

Figure 5.6. Effective surface current densities magnitude and sign of thermoelectretized polyethylene coatings vs. friction path (p = 0.1 MPa): 1, original σ_{eff} = $+2.75 \cdot 10^{-9}$ C/cm^2; 2, non-electretized specimen; 3, original σ_{eff} = $-1.25 \ 10^{-9}$ C/cm^2; 4, original σ_{eff} = $-5.8 \ 10^{-9}$ C/cm^2 .

122   TRIBOELECRET STATE OF POLYMERS

to some magnitude typical for a given friction mode, irrespective of its
sign and original magnitude. To make the residual ESCD steady, a certain
time is needed that depends on the friction conditions, while
electretization induced in advance can have quite a strong effect.

Hence, the electretization induced in advance strongly affects the wear
rate.

5.2. THE COMBINED EFFECT OF ELECTRETIZATION AND TRIBOELECTRIFICATION

The electretization induced in advance sets the vector of the electret field
intensity. The field produced by triboelectrification will amplify or
attenuate this intensity. Approximate estimates based on the postulates of
the electrical theory of molecular interactions between solids and the
molecular mechanical theory of friction allow evaluation of the force of
mutual attraction using the formula

$$F_a = k A a \sigma^2. \tag{5.2}$$

Then, it is enough to apply electretization of a polymer to change the
force of friction of polyethylene by 10% (at p=0.1 MPa) and to bring the
ESCD to about $3 \cdot 10^{-8}$ C/cm^2, as it is proved by experimental data. The
phenomenon, which is due to the superimposition of the fields of the
polymer electret (the electret state induced in advance and the tribocharge
generated in friction), is termed the superimposition by electret
triboelectrification (SET), and it especially interesting specifically when
polymer electrets rub against metals. In tests 400 μm PTFE film was
thermally electretized using a dielectric PTFE spacer at T_e=390 K, the
field tension E_e=3...600 kV/cm, the exposure time T_e in the field τ_e = 1.8
ks and cooling time in the field τ_{cool} = 0.6 ks. A roller of steel 45 40mm
in diameter and 12 mm wide acted as the counterbody. The relative
humidity was 70±5% at sliding velocity of the grounded roller v = 0.5 m/s
and the force on the polymer surface N = 20 N. The track width after 60 s
of friction was about 20 mm. A probe technique served to register the
charge distribution (the potential) over the surface of polymer specimens
before and after friction. A P2-1 instrument (the probe diameter 0.45 mm)
served for measuring the electrical parameters with the spacing between
the probe and the surface 2 mm, whereas the polymer surface moved
relative to it with the speed of about 4 mm/min. A grounded metallic

holder held the specimen on the moving table. To evaluate the charge distribution over the surface, the positive $+X$ direction was set along the direction of rotation of the metallic counterbody (Fig. 5.7 b).

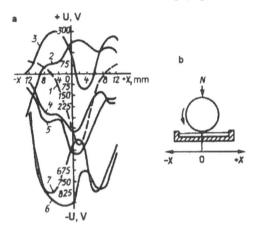

Figure 5.7.   (a) Curves of potential distribution over polytetrafluorethylene surfaces of original specimens after friction (1) and electretized before friction (2,4,6) and after friction (3,5,7). (b) Diagram of polymer-metal frictional contact and related system of coordinates.

Experimentation has demonstrated that the residual tribocharge q_f^0 is distributed over the PTFE surface asymmetrically with respect to the center of the friction path with a typical shift of the charge peak 2 mm towards the $+X$ direction (Fig. 5.7 a, curve 1). The observed shift of the negative charge peak q_f^0 can be explained by the emission of electrons from the metal when the tribocontact between the metal and the polymer is suspended, in accordance with the electrical theory of adhesion between solids [15,16].

The thermal electretization of polymer specimens produces the effective surface charge q_{eff} with its uneven distribution over the surface. The charge electrical field of the PTFE electret strongly affects the degree of the shifting of the charge peak q_Σ: the positive charge q_{eff+} increases (Fig. 5.7 a, curve 3), whereas the negative charge q_{eff-} reduces the extent of the shifting (Fig.5.7 a, curves $5,7$), also evidencing the effect of emission of electrons. It is confirmed by the results reported in [17] showing that the surface charge makes the emission more intense.

The contribution of the emission is especially remarkable when the fields of the negative charge q_{eff} of the electret are imposed on the tribocharge q_f^0 of the same polarity. In case there are no emission or gas discharges, the resulting residual charge q_Σ does not depend on the magnitude of the initial charge q_{eff}. After friction of the negatively charged electret, the charge magnitude registered experimentally is $q_\Sigma < q_f^0 + q_{eff}$ (Fig. 5.7 a, curves $1,4-7$). This is explained by the fact that the pattern of distribution of the resulting residual charge q_Σ obtained experimentally is distorted by emission and gas discharges with their contribution growing as a function of the absolute magnitude of the negative charge q_{eff}.

Emission from positively charged surfaces of polymer electrets is insignificant, therefore the experimental value of the resulting charge q_Σ is a sum of the charges of the electret q_{eff} and the tribocharge q_f^0 (Fig. 5.7 a, curves $1-3$). In the majority of cases beyond the contact area the magnitude q_Σ exceeds the initial magnitude of q_{eff}. This is apparently due to the migration of negative charge carriers and their discharge on the grounded counterbody and the holder.

Experiments with PE coatings 250–300 μm thick yielded similar results (Fig. 5.8). In case the coatings are not electretized in advance, the residual tribocharge distributes visibly symmetrically with respect to the vertical axis passing through the origin (Fig. 5.8, curve 1 for HDPE) or through some other point (Fig. 5.7, curve 1 for PTFE).

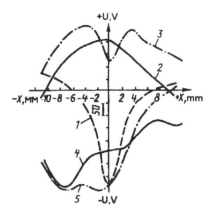

Figure 5.8. Curves of potential distribution over original high density polyethylene specimens after friction (1) and electretized before (2,4) and after friction (3,5).

The extent of the shifting is apparently caused by the dielectric properties of the studied coatings.

When the specimens are electretized in advance, the curves of the potential distribution over the surfaces before the friction are strongly different from the similar curves after the friction. When the signs of the tribocharge and the electret charge are opposite (Fig. 5.8, curves *2,3*), the potential distribution over the surface changes more than in the case of the same charge (Fig. 5.7, *a*; 5.8, curves *4,5*). Still, both these cases demonstrate cooperation between the fields of the electret charge and the tribocharge corresponding to the principle of ETS.

So, the results confirm the fact of the superimposition of the fields of the polymer electret and the appearance of the tribocharge when polymers electretized in advance and they brought into friction against metals.

5.3. THE ROLE OF ELECTRET STATE IN POLYMER COMPOSITES

Introduction of dispersed fillers into polymers increases the number of defects and the density of surface and volume states. A dispersed filler in the electret state creates a space charge in the specimen. Its field in response to the gradient effect either reduces or increases the potential barrier governing the penetration of charges generated at friction.

This mechanism of triboelectrification of the PE-based composite with the dispersed polymer in the electret state is treated in [18]. The original as well as mechanically activated PTFE served as a filler. The mechanical activation was performed in a circulation unit. The coatings were hot pressed using dielectric PTFE spacers at 433 K under pressure of 5 MPa during 30 s on the aluminum foil substrate. A direct "shaft – partial insert" friction pairs were tested to register the TSC during linear heating with the rate of 1.25 deg/min. The attached brass electrodes 14 mm in diameter were employed in the cell. The 'initial rise technique' of Garlic and Gibson [3] was employed to analyze the TSC-diagrams. The formula of Grossweiner [13] was applied to determine the trapping section and the frequency of attempts of the carriers to escape.

The results were the following. The TSC spectrum of the original HDPE coatings containing mechanically activated PTFE powder shows two peaks at T=318...320 and 393...398 K and the peak-plateau at 403...418 K. Higher concentrations of the electret filler increase the intensity of the two peaks and reduction of the peak-plateau, also changing the nature of the low temperature peak.

The TSC diagrams of the PE-based coatings containing the original and mechanically activated fluoroplastic powders manifest that after friction two peaks at 310...333 and 390...395 K and the peak-plateau at 403 K (Fig. 5.9) exist. Higher concentrations of the filler reduce the peak at 390...395 K.

The TSC data of friction between PE composites in Table 5.2 can be interpreted in the following way. In real conditions the dispersed polymers are in the electret state generated by the manufacturing conditions, including mechanical effects. The mechanical activation increases the number of structural defects, the density of the surface states with the resulting stronger electret state parameters.

Proceeding from the assumption that the dynamic contacts between the polymer and the metal are accompanied by the injection of carriers from the metal into the polymer, similar to the injection of a charge into the polymer volume, the injected and the induced charges of the same polarity would create an additional potential barrier resisting penetration of the injected charges. The reference numbers for estimation of the parameters can be those obtained for the HDPE coatings containing 25 mass % of the original PTFE (W=1.73 eV, S=8.39 10^{-11} m^2, v=3.72 10^{19} s^-) (Fig.5.9, curve 1).

Then, the greater concentration of the PTFE powder (up to 75 mass %) would apparently increase the potential barrier resisting penetration of the injected charges during friction. In fact, the greater concentration of the PTFE powder would also reduce the peak intensity at 390 K (Fig. 5.9, curve 2) together with other parameters (the activation energy, the section of the trapping of carriers and the frequency of escape efforts) (Table 5.4).

Table 5.4. Data of thermally stimulated currents diagrams processing.

Composition, mass %	T_m, K	σ_{eff}, nC/cm^2	W, eV	τ, s (at 293 K)	τ_m, s (at T_m)	v, s$^-$	S, m^2
PTFE-25, PE-75 (PTFE original)	391	−2.56	1.73	1.05· 10^{10}	365.5	3.72· 10^{19}	8.39· 10^{-11}
PTFE-75, PE-25 (PTFE original)	391	−1.13	1.11	3.48· 10^7	569.7	4.06· 10^{11}	9.16
PTFE-25, PE-75 (PTFE mechanically activated)	387	−0.44	0.87	3.07· 10^6	715.8	3.75· 10^8	8.63· 10^{-22}
PTFE-75, PE-25 (PTFE mechanically activated)	389	−0.34	0.75	1.27· 10^6	834.6	1.21· 10^7	2.76· 10^{-23}
PTFE - polytetrafluorethylene, PE - polyethylene							

Introduction of the mechanically activated PTFE in the electret state has increased the barrier even more and has reduced the intensity of the peaks and values of the parameters W, S and v (Fig. 5.9, curves $3,4$; Table 5.4).

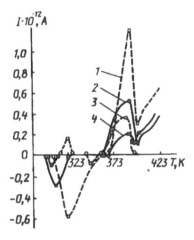

Figure 5.9. Post-friction thermally stimulated currents spectra of high density polyethylene coatings containing 25 mass % (1) and 75 mass % (2) of original polytetrafluorethylene powder; 25 mass % (3) and 75 mass % (4) of mechanically activated polytetrafluorethylene powder.

Thus, the TSC diagrams (the peak intensity at 390 K) and the table (variations of the ESCD of the coatings) indicate that higher concentrations of the filler and its introduction in the electret state reduce the residual charge. When the negatively charged PTFE filler is introduced, the energy barrier goes up decreasing the injection of charge carriers from the metallic counterbody and making the ESCD to decrease by more than seven times.

The results are also confirmed by the analysis of the type of the traps, their parameters given in Table 5.2. The trapping section S is known to change from 10^{-15} m^2 for the Coulomb centers of attraction to 10^{-25} m^2 for the Coulomb centers of repulsion. This cross-section S has the value of $2.76 \cdot 10^{-23}$ m^2 for the centers with electret fillers presumably corresponding to the Coulomb centers of repulsion which leads to reduction of the ESCD.

It is extremely difficult to identify the nature of the phenomena responsible for each individual peak. TSC peaks at 390 K are worthwhile to discuss. First, the polarity of the peak for the specimens before friction is negative, and then becomes positive after friction. Moreover, the TSC peak intensity of the specimens with the mechanically activated PTFE is much less than otherwise, and the principle of superimposition is applicable in this case. During friction of the composite with the electret filler the field of the electret adds to the field produced by triboelectrification due to inhibition of the injection of the charge carriers from the metallic counterbody. According to Figure 5.9 and the data in [18] the vector of intensity of the electret field is negative, whereas the fields produced by triboelectrification have an opposite vector resulting in a visible attenuation of the total field (Fig. 5.9, curves *3,4*).

Figure 5.10 shows the curves of variations of the force of friction confirming the above conclusion. Introduction of the electret filler (the mechanically activated PTFE powder) results in a smaller force of friction (curve *2*).

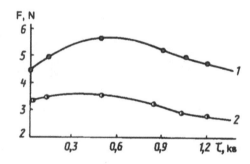

Figure 5.10. (1) Kinetics of friction force variations of high density polyethylene - based composites containing 75 mass % of non-activated powder and (2) mechanically activated polytetrafluorethylene powder.

Thus, the experimentation has indicated that the introduction of a composite filler in the electret state affects the electrification of the polymer in contact with the metal. The field appearing during triboelectrification can be amplified or attenuated in response to the vector of the field intensity produced by the particles of the electret filler according to the principle of superimposition. The free charge and the polarization are the main factors producing the electret state in traditional electrets. The   Maxwell-Wagner polarization plays a major role in heterogeneous systems which are composites and polyelectrets [19]. It is worthwhile   investigating the electrophysical   processes responsible for the appearance and   the transformation of electret systems in the polymer-polymer systems, especially in PE-PVC compositions.

The following concentrations were selected for experimentation: 100 mass % HDPE, 75 mass % HDPE + 25 mass % PVC; 50 mass % HDPE + 50 mass % PVC, 25 mass % HDPE + 75 mass % PVC, and 100 mass % PVC. The specimen coatings 300–500 μm thick were fabricated by hot pressing; the specimens were electretized in the constant electrical field (thermoelectrets) and in the crown discharge (the crown electrets). When thermoelectrets were prepared, the PTFE spacers 100 μm thick were used, and the crown electrets were fabricated at the room temperature with the voltage of the crown inducing electrode being 5 10^4 V. The spacing between the electrode and the specimen surface was 10^{-2} m, the time of the crown electrification was τ=1.5 ks. TSC and variations of the ESCD served to identify the electret state. The specimens were thermally depolarized in a temperature controlled   chamber with the linear heating rate 2.5 deg/min in the open circuit. A voltmeter registered the thermally stimulated current.

The charge magnitude acquired by the dielectric during electretization is known to be a function of the internal structure or the composition of molecules, the extent of crystallinity, and the heterogeneity. Hence it is justified to assume that the ESCD, the time of charge relaxation and the components of a composite electret mutually correlate.

Table 5.5 gives the experimental data proving that the tendency of a composite to get electrified depends upon the ESCD magnitude χ acquired by the specimen a result of electretization as the function of the tension E of the effective field: $\chi = \sigma_{eff}/E$.

Generally, the relation between the tendency to be electrified and the components of the composite has an extreme nature: the maxima are observed at the ratios between the components being 50 mass % HDPE: 50 mass % PVC and 75 mass % HDPE: 25 mass % PVC for both the crown and thermoelectrets. It should have been expected that the original polymers are electrified differently in the field and in the crown discharge:

the polar PVC is electrified quicker in the field, the non-polar HDPE- in the crown discharge (Table 5.5).

Table 5.5. Electrification ability of composites with various concentrations of components.

Composition, mass %	$\chi = \sigma_{eff}/E$, µC/(m MB)	
	electretization in field	electretization in crown charge
PVC-100	6.3	4.2
PVC-75:HDPE-25	2.3	7.0
PVC-50:HDPE-50	3.7	18.6
PVC-25:HDPE-75	3.9	21.6
HDPE-100	3.1	9.3

PVC -polyvinylchloride, HDPE -high density polyethylene

When electretized in the crown discharge, the electret charge primarily results from ionization of the surface states by the crown discharge plasma. Unlike PVC, the polyethylene has a much more "defective" structure due to the heterogeneity caused by a multitude of boundaries separating the amorphous and crystalline phases. They affect the behavior of the composite in that the structural heterogeneity and the percent concentration of the stronger electrified HDPE increase, resulting in the superimposition of the fields due to the ionization of the surface states concentrating along the boundaries separating the phases of the composite.

The increasing HDPE concentration in thermoelectrets raises their structural heterogeneity at the expense of simultaneous reduction of the concentration of the polar PVC with a greater degree of electrification in the field. Hence, the initial tendency of the composite to get electrified declines followed by the sharp heterogeneity's rise. However, the nature of the obtained dependence evidences an intricate mechanism of generation of the electret state including both the dipole polarization of PVC and the space-charge of HDPE and a definite contribution of the increased structural heterogeneity. In this case the higher concentrations of surface states at the interface boundaries both favor the Maxwell-Wagner polarization and inhibit a complete dipole polarization of the volumes containing PVC.

Kinetic studies (see [20], Fig.1) show that the curves of the ESCD of HDPE and composite electrets are satisfactorily described by the equation

$$\sigma_t = \sigma_0 - (\sigma_0 - \sigma_e)\ln t. \qquad (5.3)$$

The decay kinetics for the PVC-electrets is described by the equation

$$\sigma_t = \sigma_0 t^{-0.32}, \qquad (5.4)$$

where σ_0 is the ESCD at $t=0$; σ_e is the ESCD at $t=e$; and t is the time (hr). The ESCD relaxation time calculated using formulae (5.3) and (5.4) agrees well with the experimental figures: the experimental and the estimated figures for the HDPE-electrets are 174.0 and 174.1, respectively; and 15.0 and 16.37 hr for the PVC-electrets. The applicability of equation (5.3) for the description of the decay kinetics for PE and composite electrets evidence similar mechanisms of relaxation of the electret charge in HDPE and HDPE-containing composites and the dominating role of PE in the mechanism responsible for their "charge-confinement" effect.

The TSC spectra of the thermoelectrets studied are characterized by two peaks, one in the low-temperature range of 330...340 K and the other one in the high-temperature range of 360...370 K (Fig. 5.11). The low-temperature negative peak corresponds to the relaxation of the positive charge, whereas the high-temperature peak, on the contrary, corresponds to the negative charge relaxation. The included TSC diagrams, however, show that the low-temperature peaks of the composite electrets (curves 3–5) do not coincide with the low-temperature peaks of any component (curves 1,2). The position of high-temperature TSC peaks of the composites essentially coincides with the peaks of PVC, yet they have the opposite polarity. In contrast, their polarity coincides with that one of HDPE peaks, but their positions are noticeably different on the temperature scale.

This proves that the major contribution to the low-temperature peak comes from the dipole polarization (PVC) and the Maxwell-Wagner polarization (PVC, HDPE) and a small contribution from the space-charge polarization (PVC, HDPE). The dependence of the intensity of the TSC peaks on the ratios between the components in the composites governing the total area of interphase boundaries, allows us to conclude that the appearance of the high-temperature peaks is due to the liberation of the charges trapped by the surface states during electrification, i.e. due to the Maxwell-Wagner polarization.

An additional indication of the Maxwell-Wagner origin of the TSC peaks in the composites is the coincidence of temperatures of the TSC peaks of the thermo- and crown-electrets and similar values of the activation energy of the thermocurrent of the composite electrets fabricated by different techniques. For example, the activation energy magnitudes for the thermoelectrets containing 50 mass % of HDPE: 50

mass % PVC and 75 mass % of HDPE : 25 mass % of PVC amount to 0.77 and 0.9 V, respectively. For the crown-electrets of the same compositions these values become 0.79 and 0.96 V.

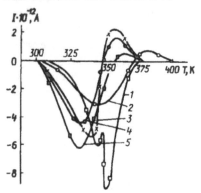

Figure 5.11. Thermally stimulated currents spectra of composite thermoelectrets: 1, polyvinylchloride (PVC); 2, high density polyethylene (HDPE); 3, 75% PVC + 25% HDPE; 4, 50% PVC + 50% HDPE; 5, 25% PVC + 75% HDPE.

Analysis of the experimental data identifies the following features of the PE – PVC electrets: similarity of the equations describing the ESCD decay kinetics of the HDPE-electrets and composite HDPE- containing electrets; presence of the peak of electrification of the material determined by the quantitative proportions of the components; the coincidence of the activation energies of the charge relaxation processes in the thermo- and crown-electrets; and the abnormal appearance of the homocharge conditioned by the structure of the material during thermal electretization using blocking electrodes.

Thus, the experimental results and the theoretical evaluation lead to the conclusion about the dominating role of the Maxwell-Wagner polarization in the generation and transformation of the electret state in the polymer-polymer composites in contrast to the space-charge dipole polarization typical for homogeneous materials.

Proceeding to the triboelectret phenomena in metal-filled composites, the available publications deal with the mechanism of conductivity of composites with metallic fillers, they outline current density leaps when the tension of the applied external field exceeds some critical level. This phenomenon is attributed either to the field emission of charge carriers or to the thermal breakthrough evolving in a composite as a factor of its composition and experimental conditions. The composites studied possess a relatively high degree of filling with a metal (over 10%) when variations of the conductivity of the metal-polymer composites are visible.

The electrophysical and mechanophysical behavior of the electrets of polymer-based composites with low filler concentrations have been sparsely studied, whereas the phenomena similar to those described above may occur in these materials. Therefore, the study has covered the electrophysical properties of the composites containing between 0.05 and 10 volume % of finely dispersed copper with isotropic distribution in the PE matrix. HDPE with the density of 0.94 g/cm^3 and the melting temperature of 403 K served to fabricate specimens. The IR and UV spectroscopy in the original HDPE have allowed detection of an insignificant quantity of fabrication impurities, such as catalysts, stabilizers, et cetera, hydroperoxide (R_3COOH) and anyl (RCH=CH) groups. Finely dispersed copper as a filler had the average size of particles equal to 2 μm, the particles mean surface area was 6.32 μm^2, and the particles shape mean factor was 0.42 (the ellipsoid of rotation).

Copper-polyethylene composites were fabricated by mixing up copper with polyethylene powder during 4 h. The obtained mixture was hot pressed under pressures of 10 MPa at 433-443 K to shape the composite material into plates 150...200 μm thick, whereas aluminum foil (the grounded electrode) 50 μm thick was pressed to one of the plate's sides. The specimens were subjected to thermoelectretization at 363 K at various external field intensities. Thermal depolarization was performed in the open circuit with the linear heating rate of 2.5 deg/min. An electrometric amplifier was employed to measure currents.

Introduction of the conducting filler into the polymer matrix increases the density of surface and volume states, on the one hand, and the specific conductivity of the polymer due to a greater number of structural defects, on the other hand. Investigation of the ESCD magnitude versus the tension of the electretizing field of the electret from pure PE and the one filled with finely dispersed copper has revealed that the homocharge appears at small field tensions, whereas the heterocharge appears at $E \geq 14...18$ MV/m.

The analysis of the TSC spectra of the composite PE-copper thermoelectrets containing 1 vol.% Cu proves that the external field intensity under 20 MV/m produces peaks on the TSC curves typical for the implemented design of thermodepolarization and homocharge relaxation. When the field intensity increases their intensity declines. When some critical magnitude E_α is reached ($E = 14$ MV/m in this case) the TSC spectrum has the relaxation peaks typical for the heterocharge. This critical intensity also depends on the filler concentration: the higher it is the less is the intensity.

Let us consider the following model of the processes evolving in the generation of the electret based on the composite with the small filler concentration to explain the observed dependence of the charge (the homo- or heterocharge) on field intensity.

It is a probable assumption that the matrix-filler contact is the injection type contact, *i.e.* when the copper-PE composite is fabricated from copper-free carriers, the latter are injected into PE. The injection occurs prior to the appearance of the blocking potential along the separating boundary (the Schottka barrier). This potential is a barrier to overcome which the carriers require some additional energy since the electron work function of copper is $\varphi_{Cu} = 4.5$ eV, and that of PE is $\varphi_{PE} = 4.75$ eV.

When the composite is placed into an external electrical field with the intensity below E_α, this barrier is somewhat lowered. Yet, it is insufficient to make the injection of carriers from the metal significant. The electret charge is primarily due to the injection from the air clearance between the PTFE spacer and the specimen surface producing the homocharge.

When the external field tension increases, the barrier reduces still more, which is manifested by the reduction of the intensity of the homocharge peak because of the compensation of the charge by the carriers overcoming the potential barrier over the PE – Cu boundary. When the external field intensity is $E > E_\alpha$ the barrier reduces so much that the emission of electrons from copper becomes noticeable. The assumption of the field emission is justified by the size (2 μm) and the shape (the shape factor is 0.4) of the filler particles enabling emission already at the potential of several hundreds volts.

The emitted carriers travel freely in the PE matrix towards the electrodes where they are confined by the traps existing in PE. Cooling in the field results in the heterocharge formation. The E_α is inversely proportional to the copper concentration in the composite, which is apparently due to the lesser spacing between the filler particles when they are more densely packed. This results in a greater field tension between the copper particles polarized in the external field reducing the potential barrier and facilitating the injection of the carriers from the filler.

Hence, the studies have revealed the existence of the critical value of the external field tension at which the charge changes its sign and to establish that E_σ is inversely proportional to the filler concentration in the composite electret.

It is worthwhile evaluating the effect of the concentration and dispersion of the metallic filler upon the parameters of the triboelectric state and the force of friction of composite coatings. Composites based on HDPE filled with the electrolytic powder of nickel were studied [21]. The nickel powder was first separated into two fractions with the particle sizes under 40 μm and over 71 μm. Techniques described in chapter 2 were employed to fabricate coatings and to carry out frictional tests to measure the ESCD. The current of thermostimulated depolarization (TSD) of triboelectretized coatings was registered at the linear heating rate of 1.25 deg/min in a heating chamber using attached brass electrodes with the diameter of 14 mm.

The electret state of the polymer composites was identified by measuring the ESCD and TSC. To exclude the effect of variations of dielectric permeability of the composite ε (when the filler concentration increases) the quantity σ_{eff}/ε was used for plotting instead of σ_{eff}. As the concentration of the filler particles increases, the ESCD of the coatings changes sharply, passing through the maximum, whose magnitude goes up as filler particles get smaller.

The sharp rise of the ESCD with the introduction of the filler can be explained by the competing effect of the growing density of the surface states representing the "confining" levels or the effect of the traps of the charge carriers and the growing conductivity of the composite. That is why a substantial increase of σ_{eff}/ε when the concentration of the filler reaches to 4 vol.% with the particle sizes under 40 μm , is due to the growing density of the surface states. The conductivities of coatings filled with the particles of these two sizes within the range of concentrations of up to $C = $ 10...12 vol.% have basically the same values.

The analysis of thermodepolarization indicates that the charges of the electret type appear during friction of metal-filled polyethylene coatings (Fig. 5.12). The TSC curve of the triboelectretized filler-free coatings has two peaks: an intensive peak at about 350 K and a relatively weak peak at 408 K, which relate to the relaxation of the injected tribocharge localized in the amorphous and crystalline portions, respectively (curve 1). This is also confirmed by the fact that highly stable traps of charge carriers in the crystalline phase of PE partially persist at the melting temperature of the polymer [22]. Introduction of a small quantity of the filler ($C = 1.6$ vol.%) increases the low-temperature peak and shifts it to higher temperatures (from 350 to 365 K), meanwhile, the high-temperature peak goes down, and the temperature at which the conductivity current appears goes up from 423 to 428 K (curve 2).

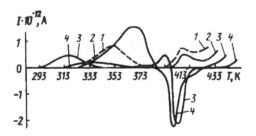

Figure 5.12. Thermally stimulated currents spectrum of triboelectretized unfilled polyethylene coatings (1) and composite coatings containing 1.6 (2), 6.3 (3) and 11.8 vol.% of nickel (4) with particle sizes under 40 μm.

These changes apparently relate to the amorphous nature of the state acquired by the polymer and to the growing concentration of the traps with higher energies in the amorphous regions [22,19]. A peak of the negative polarity current appears at $T=405$ K (and the peak at 408 K disappears) in addition to the two positive peaks at 340...343 K and 393 K in the spectrum of the TSC-coatings with 6.3 vol.% nickel (curve 3). As the filler concentration increases to 11.8 vol.% (curve 4) a low-temperature peak appears in the region of 318 K with the peak at 393 K degenerating, while the peak of the negative polarity current shifts slightly from 405 to 401 K.

It can be assumed that significant variations of the TSC current spectra at higher filler concentrations ($C=6.3$ and 11.8 vol.%) are due to the increasing concentration of traps and modification of their energy distribution as a result of the filler introduction and due to the predominant effect of deeper traps in the boundary layers of the polymer near the filler surface. Thus, the appearance of the peak of the negative-polarity current in the region of 401...405 K can be attributed to the thermal destruction of the Maxwell-Wagner polarization [19] resulting from the accumulation of charges along the polymer-filler interface. It should be noted that the TSC of a coating becomes impossible to register at higher filler concentrations ($C=14...16$ vol.%).

The frictional tests of the metal-filled PE-based composites have revealed that the relation between the friction coefficient f and the filler concentration C has an extreme pattern. The f–C dependence when one uses the filling with finely dispersed particles (under 40 μm) is characterized by the presence of the maximum, whereas for larger particles (over 70 μm) this dependence exhibits the minimum. A certain correlation is observed in the relationship σ_{eff}/ε–C relation : when the filler particles are smaller than 40 μm, the dependencies σ_{eff}/ε–C and f–C have a similar patterns, whereas for the particles larger than 70 μm, these dependencies differ significantly. The extreme dependencies of the ESCD and f upon concentrations are manifested at basically similar filler concentrations of C=3...5 vol.%. The analysis of the experimental dependencies allows us to conclude that the observed abnormal variations of the friction coefficients relate directly to both the growth and then to the decline of the generated charge magnitude, and to the alteration of the composite conductivity mechanism due to more compact arrangement of the filler particles. Greater concentrations of the particles cause the filler dispersion to affect the microgeometry of the surface and the mechanical behavior of the coating materials.

Thus, the thermodepolarization analysis has confirmed the manifestation of the triboelectret effect in the metal-filled polymer composites. The pattern of variations of the TSC spectrum of the triboelectretized HDPE coatings has been established when a metallic filler is introduced into a polymer. An abnormal growth of the magnitude of the residual tribocharge has been detected and its relation to the friction coefficient of a metal filled composite is demonstrated.

5.4. FRICTIONAL BEHAVIOR OF RADIOELECTRET POLYMER COATINGS

Electron bombardment of polymer materials is accompanied by both radiation and chemical effects and the appearance of the space homocharge, i.e. the evolution of the electret state. At present such bombardment with a nonpenetrating flux of electrons is one of the most effective techniques of generating electret (radioelectret) states of polymer dielectrics. Yet, in the studies of modifications of the frictional behavior of polymers under the effect of electron fluxes the electret state would either be ignored completely or excluded beforehand. Particularly, Silin and Dukhovskoy, in order to exclude the accumulation of space charges in the surface layers, have bombarded PTFE surface rubbing in the vacuum with a flux of electrons penetrating to the depth of 0.2...0.3 μm producing significant alteration of the friction coefficient [23].

The observed effects have not yet been explained completely. Hence, the objective has been to investigate the effect of bombardment by a monoenergetic flux of accelerated electrons of the surfaces of PE and polytetrafluorchlorethylene (PTFCE) coatings and PTFE films to reveal modifications of their electrophysical and frictional characteristics. The choice of these polymers is explained by their frequent use as antifrictional materials in tribology, their "attractive" dielectric behavior and their differences in irreversible radiation effects: PE is predominantly a bonding polymer, PTFE and PTFCE are predominantly destructive polymers.

The specimens of the HDPE and PTFCE coatings were fabricated on the aluminum foil 100 μm thick using the technique described in [24]. The PTFCE coatings were produced at 540 K and they were 360...420 μm thick. Industrial PTFCE film 400 μm thick was also tested. The polymer surface was bombarded in the vacuum (6.7· 10^{-3} Pa) with a flux of electrons having the energy E=75 keV and the current density i=1.02 10^{-2} A/m². To reduce radiation electrical conductivity after bombardment the specimens of the PE and PTFCE coatings were heat-treated during 1.8 ks at 343 K and PTFE films at 363 K. The ESCD was estimated using a noncontact compensating technique with a vibrating electrode. Friction tests were set-up as a shaft and a partial insert in a friction apparatus Mi–1M. Steel 45 rollers with the initial roughness R_a acted as counterbodies. The wear rate was determined by weighing after 21.6; 7.2 and 3.6 ks of continuous operation for PE, PTFCE coatings and PTFE films. The temperature of the specimens was checked using a thermocouple installed between the substrate of the coating and the surface of the metallic insert to which the specimen was attached. The microhardness of the polymer specimens was estimated with a "Neophot" instrument under the load of 0.2 N for the PE and PTFCE coatings and 0.1 N for PTFE films. Radiographic structural analysis and IR-spectroscopy with the help of the DRON–2 apparatus and a UR–20 spectrometer were employed to monitor the structural and chemophysical modifications evolving in the polymers during electron bombardment.

Table 5.6 lists the wear rates and the linear wear of original coatings during tests. It has been established experimentally that electron bombardment (radioelectretization) would strongly alter the frictional characteristics of the polymers studied. The tests of the PE coatings have revealed that prolonged electron bombardment produces a fast rise of the friction coefficient and the temperature of a specimen which have basically constant values for $t > 0.3$ ks. The wear rate of the PE coatings reduces by more than 10 times already after $t\sim180$ s of bombardment, and when t is increased further it rises somewhat and then reaches the minimum at $t\sim1.2$ ks. The ESCD of the PE coatings one day after radioelectretization changes extremely in response to the duration of bombardment reaching the maximum approximately after $t = 0.2...0.35$ ks when the wear rate reduction is observed.

Table 5.6. Wear rates l_0 and linear wear values of original coatings during test time t.

Coating material	Friction conditions			$l_0 \cdot 10^{10}$	Wear, μm
	P, MPa	v, m/s	t, ks		
PE	0.5	0.5	21.6	7.29	7.9
PTFCE	0.25	0.5	7.2	46.40	16.7
PTFE	0.1	0.5	3.6	485.00	87.3

PE - polyethylene, PTFCE - polytetrafluorchlorethylene, PTFE - polytetrafluorethylene

The friction coefficient of the PTFCE coatings is not affected by the duration of friction, whereas ESCD changing extremely passing through the maximum. Like the PE coatings, the maximum ESCD of bombarded PTFCE coatings is reached within the t range in which the wear rate reduction is observed. Evaluation of the variations of the ESCD and the wear rate in response to the duration of electron bombardment of the PE and PTFCE coatings allows us to conclude that one of the causes of such higher wear resistance of radioelectretized polymer materials is the space charge field effect upon the intensity of dispersion of the polymer surface layer and the frictional transfer.

To investigate the effect of the space charge field on the frictional behavior of the nonpolar polymer (PTFE film) the two sides were scrutinized: the side bombarded with electrons and the opposite unbombarded side. The results prove that the wear rate of the bombarded side of the PTFE film reduces sharply in response to the duration of electron bombardment, reaching the minimum after $t=0.3...0.5$ ks (see Table 5.4), and for $t > 0.7$ ks the wear rate increases again. The wear rate of the unbombarded side reduces monotonously as a function of t and coincides with the rates typical for PTFE films thermally electretized at 400 K; no significant variations of the friction coefficient of the PTFE film have been registered. The ESCD of the two sides of the film increases monotonously as a function of radioelectretization duration. Therefore, the effect of the space charge field not localized directly in the wearing surface layer on the frictional behavior of PTFE films is insignificant.

During bombardment the primary flux of accelerated electrons is decelerated in the polymer and produces a space charge in a fine surface layer. This flux has the maximum density at a depth equal to its average travel distance R which depends on the energy of electrons E and the absorption properties of the polymer. For example, $E=75$ keV yields the magnitudes of R approximately equal to 75 and 42 µm for PE and PTFE, respectively. The most essential irreversible processes after bombardment are the appearance of the spatial bonded structure and the destruction of the polymer macromolecules. Based on the known relations, the estimated mean doses [19] absorbed by the polymer layer in response to the bombardment duration evidence that significant structural and chemophysical modifications evolve in the materials studied that are the leading contributors into the alterations of the frictional behavior. The radiographic structural analysis and IR-spectroscopy prove this because the bombardment increases the crystallinity of the polymer specimens (Table 5.7), and the concentration of double bonds, carboxyl, and terminal groups in the chains of macromolecules.

Table 5.7. Variations of polymer materials crystallinity, % in response to duration of electron bombardment.

Material	Duration of bombardment, ks			
	0	0.18	0.3	1.2
PE	60	65	63	65
PTFCE	51	60	65	62
PTFE	68	73	82	82

PE - polyethylene, PTFCE - polytetrafluorchlorethylene, PTFE - polytetrafluorethylene

The significance of bombardment is also shown by a stronger microhardness, as the structural-responsive characteristic and other data demonstrate the appearance of the spatial bonded structures in the surface layers at the PE coatings. It is demonstrated [25] that the dose of radiation equal to 100 Mrad brings the concentration of the gel-fraction in the PE specimens to about 85% and 200 Mrad to 90%. Hence, the observed pattern of modification of the frictional characteristics of the PE coatings can be attributed to the hardening of the surface layers produced by bombardment.

Reduction of the microhardness of PTFCE coatings upon bombardment proves the impairment of the physical and mechanical properties of the surface layers of polymers due to the destruction of macromolecules. Prolonged bombardment of PTFCE coatings reduces their wear resistance. Bombardment of the PTFE films increases the crystallinity of the polymer surface layer by "compacting" the macromolecules. Due to the reduction of the molecular mobility the brittleness somewhat increases apparently leading to a stronger microhardness and a reduced wear rate of the bombarded side. The microhardness of the unbombarded side of the PTFE specimens remains basically unchanged. A stronger wear rate after prolonged bombardment (t=1.2 ks, D=1000 Mrad) is caused by the impaired mechanical properties and the embrittlement of the surface layer of a polymer, as it is known to happen for the PTFE after considerable radiation doses.

Thus, the bombardment of polymer coatings with nonpenetrating electrons accompanied by the appearance of the radioelectret state, leads to significant modifications of the frictional characteristics of coatings. They are basically caused by radiation, chemophysical and structural transformations of a polymer. The tested electron bombardment modes (high absorbed doses) do not produce any significant contribution of the space charge to the modifications of the frictional behavior of coatings compared with radiation and chemical effects. The results prove the effectiveness of application of optimum doses of electron bombardment for improving the wear resistance of the surface layers of polymers.

5.5. MECHANISM OF PRODUCING THE ELECTRET STATE IN FRICTION OF POLYMERS

Application of the basic postulates of the electrical theory of adhesion to the explanation of the characteristics of electrification of polymers in friction and conclusions of the electron theory of disordered systems and surface states have verified the mechanism of electrification. The presence of surface states that attract of electrons injected from the metallic counterbody when the contact is broken produce an electrical charge on the surface of the polymer specimen. The triboelectrical field of the surface charge produces charge redistribution in the surface region yielding the spatial charge . A double electrical layer appears along the polymer – metal boundary. The charge density and the structure of the layer are governed by the characteristics of the surface states.

The mechanism of electretization of polymers in friction seems to be as follows: during electrification a space charge and a spatial charge region appear causing polarization of a polymer material in the field and (with the help of injected charges) an electret state is generated. The parameters of the state are conditioned by the free injected charge carriers as well as the polarization. The triboelectret state significantly affects the frictional characteristics of polymers.

REFERENCES

1. Guzenkov S.I.. *Soviet Journal of Friction and Wear*, vol. 11, no 1, pp. 151-154, 1990.
2. Belyi V.A., A.F. Klimovich, V.S. Mironov. *Proc. of the BSSR Academy of Science*, vol. 26, no 1, pp. 39-42, 1982.
3. Lushchejkin G.A.. *Polymer Electrets* (in Russian), Moscow, 1976.
4. Bartenev G.M.. *Strength and Mechanism of Failure of Polymers* (in Russian), Moscow, 1984.
5. Bartenev G.M., and Yu.V. Zelenev. *Course in Physics of Polymers*(in Russian), Leningrad, 1972.
6. Sviridenok A.I.. *Tribology Int.*, vol. 24, no 1, pp. 37-44, February 1991.
7. *Polymers in Friction Units of Machines and Instruments/* Ed. by A.V. Chichinadze (in Russian), Moscow, 1988.
8. Margis D.. *J.Mater.Sci.*, vol. 20, pp. 3041-73, 1985.
9. Briscoe B.J., K.Fridrich (ed.).*Friction and Wear of Polymer Composites*. Elsevier, Amsterdam, Ch.2,p.25, 1986.
10. Klimovich A.F., and V.S. Mironov. *Soviet Journal of Friction and Wear*, vol. 2, no 4, pp. 113-117, 1981.
11. Kostetskii B.N., M.G. Nosovskii, and L.I. Bershadskii. *Surface Strength of Materials in Friction* (in Russian), Kiev, 1976.
12. Gershman I.S., and N.A. Bushe. *Journal of Friction and Wear*, vol. 16, no 1, pp. 41-48, 1995.
13. Harper W.R.. *Contact and Frictional Electrification*, Oxford, 371 pp., 1967.
14. Belyi V.A., A.I. Sviridenok, M.I. Petrokovets, and V.G. Savkin. *Friction and Wear in Polymer-Based Material*. Pergamon Press, N.Y., 1982.
15. Deryagin B.V., N.A. Krotova, and Yu.A. Khrustalev. *Adhesion of Solids*. Moscow, 1973.
16. Balachandran W.. *Tribology in Particulate Technology*. Ed. by B.J. Briscoe and M.J. Adams. Adam Hilgen, pp. 135-154, 1987.
17. Vallbrandt I., U. Brjukner, and E. Linke. *Proc. of Symposium on Mechanochemistry and Mechanoemission of Solids* (in Russian), Tallinn, pp. 46-47, 1981.
18. Guzenkov S.I., Yu.V. Gromyko, and A.F. Klimovich. *Soviet Journal of Friction and Wear*, vol. 8, no 1, pp. 107–110, 1987.
19. *Electrets* (English translation), Moscow, 1983.
20. Gromyko Yu.V., and A.F. Klimovich. *Proc. of BSSR Acad. of Sci.*, vol. XXXIII, no 6, pp. 531-534, 1989.

21. Mironov V.S., and A.F. Klimovich. *Proc. of BSSR Acad. of Sci.,* vol. 30, no 8, pp. 724-727, 1986.

22. Vannikov A.V., V.K. Matveev, V.P. Sichkarev, and A.P. Tyutnev. *Radiation Effects in Polymers.Electrical Properties,* (in Russian), Moscow, 1982.

23. Silin A.A.. *Friction and Its Role in The Progress of Technology*(in Russian), 176 pp., Moscow, 1983.

24. Klimovich A.F., and V.S. Mironov. *Soviet Journal of Friction and Wear,* vol. 2, no 4, pp. 113-117, 1981.

25. Pleskachevskii Yu.M., V.V. Smirnov, and V.M. Makarenko. *Introduction into Radiation Science of Polymer Composites* (in Russian), 191 pp., Minsk, 1991.

CHAPTER 6

ENVIRONMENTAL EFFECTS ON ELECTROPHYSICAL PROCESSES IN FRICTION

The processes of generation, accumulation, and relaxation of electrical charges on surfaces and in the bulk of polymers are governed by a variety of factors: the nature and mode of friction, electrophysical properties of contacting bodies, external electrical fields, and dissipation processes (electrical conductivity, electron emission, desorption of gas discharge ions, etc.).

Accumulation and relaxation of electrical charge is significantly influenced by the contact environment. Numerous studies of triboelectrification of polymers in uncontrollable environment, in the air, as a rule, have been accomplished. Still, application of polymers and polymer composites in friction units operating in oils, water emulsions and other industrial media, in the areas with various climatic conditions necessitate studies of the electrophysical processes in friction in response to environmental factors, especially to high levels of atmospheric humidity, presence of liquids, and vacuum conditions.

6.1. EFFECT OF ATMOSPHERIC HUMIDITY

The effect of atmospheric humidity on the electrophysical processes in the friction of polymers manifests itself both at the stages of electrification, generation of the electret state and during subsequent operation and electrical charge relaxation. The role of the relative air humidity has been described in detail by many publications.

When the atmospheric humidity increases, the majority of electrets are known to lose their stability. Partial restoration of the charge magnitude may occur when the humidity reduces [1].The authors of [2] divide the processes of the charge decay in a humid environment into reversible and irreversible ones . Charged particles of the opposite sign concentrating in the drops of adsorbed moisture screen the electret charge and reverse the charge decay. Injection of charged particles of the opposite sign and their trapping produce an irreversible charge decay.

The curves of charge relaxation in various electrets produced from Teflon FEP (Fig. 6.1) exposed to different temperatures and degrees of environmental humidities evidence primarily that Teflon preserves the charge under normal environmental conditions, while manifesting mild charge relaxation at high temperatures and humidity. Strong charge stability of electrets charged with electron beams compared with thermoelectrets at elevated humidity is explained [3] by the protection of deeply trapped charges against atmosphere. In this connection Cessler [4] concludes that the electret external electrical field also attracts polar particles, for example, water molecules. These molecules do not cause external relaxation because a complete discharge does not exist, yet they frequently assists the acceleration of the internal relaxation processes.

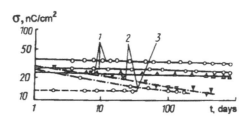

Figure 6.1. Charge relaxation of various electrets from Teflon FEP at different temperatures and humidities: *1* — thermoelectret at, 22 °C and 40% humidity; *2* — at 70 °C and 100% humidity; *3* — electret generated by electron flux at 70 °C and 100% humidity

The effect of the adsorbed moisture on the reduction of both surface conductivity as well as the effect of the magnitude of the emission flux on the destruction of dielectrics has been shown [5]. A layer of moisture on the surface may be a cause of the electrolytic mechanism of electretization during friction of solids, as Loeb has remarked [6]. The water film reacts with a solid by exchanging ions until the equilibrium is achieved. Friction may remove the film, but the body will still remain charged. Bowden and

Davy note a possibility of this mechanism. Lowell [7] has indicated the lack of experimental data proving the involvement of this mechanism in triboelectrification of polymers.

Humidity of the atmospheric air significantly influences the magnitude of the surface conductivity of dielectrics because the adsorbed water film usually contains a great quantity of ion from impurities and dissolved substance. Other conditions being equal, the conductivity of the film is determined by its thickness, and, hence, it depends much upon the relative humidity: the higher it is the thicker the film. When humidity is high the surface conductivity becomes the decisive factor leading to strong current leaks, and, therefore, to the reduction or suppression of electrification of polymers [8,9].

Shashoua [10] investigated the relation between the antistatic behavior of chemically structured polymers and has demonstrated that significant water absorption or even polymer solubility in water should not cause any abrupt change in the charge leak. Therefore, polymers have different leak rates for positive and negative charges. This indicates that the charge leak is not possible only due to conductivity, because in this case similar leak rates for both positive and negative charges should be expected.

Humidity significantly affects generation and relaxation of the electret states in 'classical' electrets as well as accumulation of charges in polyethylene when external electrical field is applied [11]. Two additional peaks have been discovered in the thermally stimulated currents (TSC) spectrum of moisture-saturated polyethylene (PE) specimens because of water adsorption. It was demonstrated that the additional peaks are caused by the liberation of free charge carriers, not by dipole polarization. The additional peak in the TSC spectrum of the PE specimens kept in a humid atmosphere before polarization was detected by the authors of [12]. It is attributed to the relaxation of ions resulting from ionization of adsorbed water molecules in the space charge field trapped along the boundary separating the amorphous and the crystalline regions. There are interesting studies of the effect of water adsorption on the generation of the electret state in the thermoelectrets of the oxidized high density polyethylene (HDPE) that have demonstrated the possibilities of the appearance of the abnormal discharge current [13]. The main reactions governing the chemical transformations in the crown discharge in the moist atmosphere noticeably affecting the generation of the electret state in the crown electrets of PE and polypropylene (PP) are described in [14].

The charge surface density of an electret has a strong non-linear dependence on the air humidity. The moisture effect has a reversible nature: whereas the surface charge density declines basically to zero at relative humidity >90%, as it becomes drier its magnitude restores to almost the initial level. It is assumed that the variations of the charge surface density are due to the "short-circuiting" effect of the moisture film condensing on the charged surfaces [15].

The relation between the original charge density and the applied Maylar films of polyethylene terephtalate (PETP) is shown in Fig. 6.2. The charge density shows some increase when the air is humid during electrification. There are interesting studies of the effect of the air humidity on the kinetics of triboelectrification and the generation of the electret state in polymers [16]. The study included polymers with different degrees of water absorption, a hydrophobic HDPE with $W =$ $(0.15...0.3) \times 10^{-10}$ g/cm h Pa) and a hydrophilic polycaproamide (PCA) with $W = (3...6) \, 10^{-10}$ g/(cm h Pa).

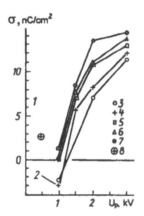

Figure 6.2. Initial charge density vs. applied tension of polyethylene terephtalate films 180 μm thick electrified in different gases: 1 — homocharge; 2 — heterocharge; 3 — CO_2(dry); 4 — O_2 (dry); 5 — air(dry); 6 — air(moist); 7 — N_2(dry); 8 — H(dry)

The coatings were made on aluminum foil of type A99 0.1 mm thick type A99 by hot pressing using dielectric polytetrafluorethylene (PTFE) spacers. Frictional contact of the polymer coating with the metallic counterbody was arranged in direct or reverse friction couple configured as shaft-partial insert on a stand with synchronous registration of the moment of friction and current characteristics of the process. The design of the chamber allowed to monitor, adjust and maintain a specified level of the relative air humidity.

The extent of triboelectrification was estimated based on the magnitude of the effective surface charge density using the technique from [3] in a chamber with controlled heater with linear heating of specimens with the rate of 2.5 deg/min. Blocking aluminum electrodes were employed to measure thermally stimulated currents.

Figure 6.3 shows the kinetics of the electrification current in the reversed PCA-metal couple at various relative environmental humidities χ. The figure shows that the growing air humidity extends the time until the first inversion (reversal of the electrification current), and no inversion occurs at $\chi.>60\%$ humidity.

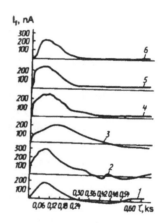

Figure 6.3. Kinetics of variations of triboelectrification current in reversed polycaproamide-metal friction couple at sliding velocity $v = 1$ m/s, nominal pressure $p = 0.15$ MPa and different air humidities, %: 1 — 30; 2 — 40; 3 — 50; 4 — 60; 5 — 70; 6 — 90

The following formula has been advanced to evaluate the triboelectrification currents (for the case when $W_m > W_n$, where W_m, W_n are the electron work function for the metal and the polymer, respectively):

$$I_f = I_c - I_m \pm I_d + I_t - I_p \qquad (6.1)$$

Here I_c is the current appearing due to the contact electrification; I_m is the current generated by the motion of the charged particles during mass transfer; I_d is the current generated by sorption and desorption in the regions underlying the contact surface; I_t is the thermocurrent; and I_p is the discharge current generated when the frictional contact is lost.

Variations of the relative air humidity (at pV = const) differently affect the different components of the electrification current. Higher humidity most strongly affects the current I_d generated by sorption and desorption. Water adsorption is known to reduce the contact difference of potentials lowering the charge transfer, and, hence, the contact electrification current. The adsorbed moisture results in a greater number of charge carriers, quicker molecular motion in the surface underlying region (plasticization) and in the intense charge dissipation, i.e. in the rise of I_p current.

Equation (6.1) shows that the first current inversion at a constant friction mode (pV = const) and humidity variation (χ = var) results from the change of the current I_d polarity, so that the remaining components of current I_f are unable to change their direction as χ increases.

Thus, the time until the first inversion with the growing relative air humidity proves the assumption that the first electrification current inversion is caused by the water desorption from the surface layers of the polymer [17].

Let us consider the effect of humidity under these conditions on the electret state based on the effective surface charge density (ESCD) magnitude, the time spectrum of the tribocharge relaxation and the TSC spectra.

Figure 6.4 shows the relations between the ESCD, the force of friction and the relative air humidity in the frictional contact between a direct friction couple of the HDPE-coatings and the metallic counterbody.

The curves have a characteristic pattern with the peaks in the range χ = 50...60% . It has been demonstrated [17,18] that the full surface charge generated during friction of a polymer corresponds to the sum of charges in the fast and the slow states, the electret state being primarily generated by the charge of slow surface states. The latter are known to be governed most frequently by adsorption and desorption processes. Thus, the growing humidity increases the density of the slow surface states of the polymer. Moreover, when the relative air humidity is taken into account, the thickness of the adsorbed moisture film increases. It leads to the greater contribution of the electrolytic mechanism of the process of electrification (in addition to the contact mechanism). Electrolysis becomes more evident when the film is about 100 nm thick. Loeb [6] shows that such layers appear on the surfaces of solids at χ =50...60%. When the film is thicker than 100 nm the surface conductivity increases strongly, whereas the dissipation of charges and the total electrification intensity decline.

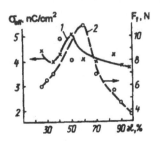

Figure 6.4. Effective surface charge density magnitude (1) and force of friction (2) of high density polyethylene coatings vs. relative air humidity (friction mode p = 0.15 MPa, v = 1.0 m/s)

Hence, the extreme dependence of the ESCD magnitude on the relative air humidity is caused, on the one hand, by the ESCD growth due to a greater density of the surface states and an additional contribution of the electrolytic mechanism of electrification, and the ESCD reduction, on the other hand, at χ >50...60% due to a greater surface conductivity and charge dissipation.

An assumption that a higher air humidity involves the mechanism of contact and electrolytic electrification in the charging of polymers in the frictional contact with a metal is confirmed by the electret thermal analysis. Figure 6.5 shows the TSC spectra of HDPE coatings after friction on the metallic counterbody at different relative air humidities.

The analysis of the TSC spectra indicates that the higher humidity creates an additional peak (II) at T = 390...393 K in addition to the low-temperature peak (I) in the region T_I=373...383 K and the high-temperature peak (III) in the region T_{III}=403...408 K. A similar peak has been found for the PE specimens exposed to humid air prior to treatment in the electrical field [12]. Appearance of this additional peak is attributed to the relaxation of ions trapped on the surface separating the crystalline and amorphous phases.

Appearance of peaks I and III typical for HDPE triboelectrets can be explained by the relaxation of charge carriers injected from the metal

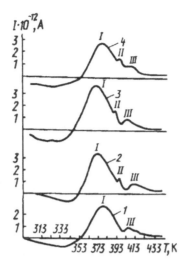

Figure 6.5. Thermally stimulated currents spectra of high density polyethylene coatings after friction at various relative air humidities, %: *1* — 30; *2*–50; *3* — 85; *4* — 99

during contact electrification and trapped along the boundaries separating the amorphous and crystalline phases (peak I) and in crystallites (peak III).

The polarity and the position of peak II manifested in the TSC spectrum at higher humidity apparently evidence the fact of relaxation of the negative ions localized in the confining centers. During the frictional contact with a metal, the polymer injects the charge carriers localized in the surface states. When the humidity of air is elevated in the space charge field due to the injected charges the adsorbed water molecules are ionized and generate the ions entering the corresponding traps. Therefore, the additional peak in the TSC spectrum after friction at elevated humidity is due to the relaxation of the ions generated in the contact zone and localized in the trapping centers. This is the result of the electrolytic mechanism of electrification due to a greater thickness of the adsorbed water film.

Thus, in the case of elevated humidity the electret state in polymers at friction on metals results from injection processes generated by the mechanism of contact and electrolytic electrification.

6.2. FRICTION IN FLUIDS

Information on the static electrification and electret effects in fluids, especially in the electrically conductive fluids, is far from exhaustive. It was demonstrated [21] that the bioelectret effect exists in a semiconducting environment of biological tissues possessing predominantly the specific resistance of $10^2...10^4$ Ohm m in water near the vegetating tissue.

A number of publications is devoted to the effect of fluids and gases of various nature on the electret charge relaxation. It is noted that the storage at elevated humidity significantly impairs the stability of electrets. A study of the effect of dielectric fluid upon the electret behavior of F-4 and PETP has revealed that the charge relaxation is determined by the surface contact between fluids and the electret [19].

Let us scrutinize the electrophysical phenomena during the friction of polymer coatings (HDPE) produced by hot pressing in liquid dielectrics and electrolytes. The fluids used in experiment were Vaseline and castor oils, glycerin, distilled water (bidistillate) and electrolytes - water 1% and 0.1% KCl solutions. The tests were set-up as 'shaft-partial insert' in a special chamber. Rollers from steel 45 with the original roughness R_a = 0.63...0.40 μm acted as the counterbodies. Frictional contacts with the metallic counterbody were used at various speeds in vessels filled with the liquids. The extent of triboelectrification was rated based on the residual ESCD and the TSC technique.

It has been noted earlier that the major criteria of the electret effect manifestations are the following : a charge large enough to generate the external electrical field, prolonged existence of the charge in tie and the thermally stimulated depolarization spectrum (TSC). The ESCD was registered in order to determine the charge of a specimen generated by the external electrical field. The ESCD dependence in the general case demonstrates the presence of a peak (Fig. 6.6) conditioned by the double effect of the accelerated parameter effect upon the kinetics of electrification. When the friction velocity is increased the charge leak and the contact interruption reduce. On the other hand, the higher velocities increase heat liberation in the contact zone and surface and the volume electrical conductivity of the polymer and the medium, hence resulting in a stronger charge leak. Friction in the air manifests a similar dependence of the ESCD on the velocity. Different mechanisms of electrification in the air and in fluids should also be noted: transfer of charge carriers in fluids during triboelectrification proceeds through the contact between juvenile polymer-metal surfaces and through the lubricating layer (the liquid medium). In this case the ESCD magnitude depends upon the electrical conductivity of the medium.

The maximum ESCD magnitude during the friction in distilled water (Fig. 6.6) is reached because the transfer of charge carriers proceeds through juvenile surfaces and through the liquid layer to some extent. Reduction of the ESCD during friction in 0.1% and 1% KCl solutions is due to strong rise of the leaking currents because of a good conductivity of the solutions (Table 6.1).

Vaseline oil generates the maximum ESCD. The medical Vaseline oil belongs to nonpolar dielectrics with the symmetric molecular structure, whereas the castor oil and the glycerin belong to the polar substances with the asymmetric molecular structure, which are known to have the best adsorption in contact with metals [21]. The better the adsorption, the larger is the work of adhesion [20]. When polymers operate in polar media producing strong oriented boundary layers on the metallic counterbodies, a direct polymer-metal contact is inhibited, and the charge carriers can be transferred through singular discontinuities of the boundary layer. It can explain lower ESCD magnitudes for glycerin and castor oil compared to Vaseline oil.

Table 6.1. Characteristics of fluids

Fluid	Polarity	Specific volumetric electrical resistance ρ_v, Ohm m	Work of adhesion W^*10^{-3} J/m^2,	Effective surface charge density ** σ_{eff}, mcC/m^2 v=0.5...2 m/s	v=0.7 m/s
Water (bidistillate)	polar	10^4	42.12	11.0	5.77
0.1 and KCl	—	1	47.91	2.20	2.16
1 and KCl	—	10^{-1}	51.83	1.60	1.05
Vaseline oil	non-polar	10^{10}	36.00	4.20	4.16
Glycerin	polar	10^6	45.14	1.25	0.94
Castor oil		10^8	predominant***	0.80	0.60

Note. *Experimentally estimated according to [20]; determined experimentally (Fig.6.6); *** W_a was estimated based on the liquid drop moving angle (from the end to the front); the castor oil has not manifested movements of drops from the back due to high value of the "escape" work (from a solid surface) proportional to W_a

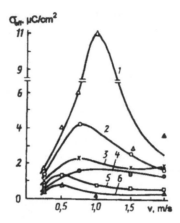

Figure 6.6. Effective surface charge density vs. friction velocities of high density polyethylene coatings in liquid dielectrics and electrolytes: *1* - water; *2* — Vaseline oil; *3*—KCl, 0.1 % solution; *4* — KCl, 1 % solution, *5* — glycerin; *6* — castor oil.

Variations of the tribocharge magnitude as a function of time have also been investigated. The charge was measured immediately after friction (σ_{ef}^1), 120 days later in the air atmosphere at the room temperature (σ_{ef}^2) and after the exposure in the desiccator for 6 days (σ_{ef}^3), and subsequent storage in the air at the room temperature during 16 days, respectively. The total storage time was 142 days, (σ_{ef}^4), the total storage time was 147 days, (σ_{ef}^5), the total storage time was 160 days, (σ_{ef}^6). The relative air humidity during storage and charge variations was 33...36%. The results indicate (Table 6.2) that the specimens triboelectretized in dielectric media and in electrolytes preserve the electrical charge for the long time (up to 160 days of tests). The charge increase is noticeable, especially after exposure in the desiccator with a subsequent gradual reduction very typical for the electret state.

One of the main characteristics of the electret state is the TSC spectrum. The thermopolarized analysis of the HDPE coatings tribocharged in dielectric and electrically conductive media confirm the electret nature of the charge (Fig. 6.7). This proves that there exists the maximum TSC current of the positive and the negative polarity.

Table 6.2. Effective surface charge density variations σ_{ef}, $\mu C/m^2$ of triboelectretized high density polyethylene specimens depending upon storage time (p = 0.1 MPa; v = 1.0 m/s; τ =1.2 ks)

Medium	Specimen thickness, μm	σ_{ef}^1	σ_{ef}^2	σ_{ef}^3	σ_{ef}^4	σ_{ef}^5	σ_{ef}^6
Vaseline oil	490	-1.3	-1.1	-2.9	-8.5	—	—
Vaseline oil	590	-2.0	-3.1	-6.0	-4.6	—	—
Vaseline oil	530	-1.2	-0.3	-0.8	-2.3	—	—
Vaseline oil	520	-1.3	-1.2	-1.0	-2.4	—	—
0.1 n KCl solution	270	-1.6	-6.1	-2.6	—	-3.1	—
0.1 n KCl solution	320	-1.3	-5.2	-2.8	—	-4.8	—
0.1 n KCl solution	220	-0.6	-0.5	-0.8	—	-0.7	—
1 n KCl solution	420	—	-0.7	-0.7	—	-5.1	—
1 n KCl solution	390	-1.2	-3.7	-1.8	—	-1.3	—
Water	580	-15.8	-3.7	-1.6	—	—	-1.9
Water	700	-4.2	-1.8	-2.3	—	—	-2.1

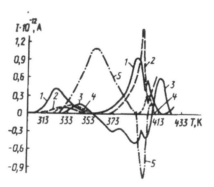

Fig. 6.7. Thermally stimulated currents current spectrum of high density polyethylene coatings triboelectretized in water (1); 0.1% KCl solution (2); Vaseline oil (3); water, after 160 of storage in the air (4), (5) the friction mode: p = 0.1 MPa, v = 1 m/s, τ =1.2 ks

The shown TSC diagrams allow identification two regions of the low-temperature (320...373 K) and high-temperature (400...420 K) peaks. It should be noted that the friction in fluids compared with friction in the air manifests the shift of the low-temperature peak towards lower temperatures. This results from the effect of the environment upon the processes of plasticization, electrification, and electretization of the polymer.

Thus, generation of the charge of the order of 1...10 mcC/m^2 in the specimens after friction in fluids, prolonged maintenance of the charge (up to 160 days and longer) and the TSC spectrum confirm the manifestation of the electret effect during friction of polymers in dielectric fluids and electrolytes. Broad application of polymers in friction units operating in oils, water emulsions, and other industrial media necessitate consideration of the established characteristics significantly affecting the electrophysical and tribological behavior of polymeric materials.

6.3. FRICTION IN VACUUM

It is especially interesting is to investigate the electrophysical phenomena accompanying the friction of polymers in the vacuum conditions when neutralization of charges is inhibited. Electrification of polymers should proceed more intensely creating strong electrical fields. When dielectrics undergo friction in the vacuum (10^{-3} Pa) the field tension may reach 10^6 V/cm activating a new mechanism of electrical charge dissipation, such as the emission of high energy electrons [22]. The vacuum frictional contact between dielectrics can be accompanied by the phenomena of luminescence, and radiographic emission.

The Metal-Polymer Research Institute (Grodno, Balarus) has accomplished an investigation of friction in the vacuum for various polymers (HDPE, PCA, PETP). They were desiccated in advance in the vacuum ($P_v = 10^{-3}$Pa) at $T = 383...393$ K during $\tau = 4$ hours. The frictional contact with metal was established in the vacuum in the reversed friction couple set-up as the "shaft-partial insert" on a specially designed stand with an attachment for mass-spectrometry measurements.

The extent of electrification of the specimens in friction was rated based on the triboelectrifiction current registered with the help of an autocompensating microvoltnanoampmeter. Mass spectrometer was employed for the qualitative analysis of volatile products released during the friction of polymeric materials in the vacuum.

Figure 6.8 shows the kinetics of HDPE electrification in friction on metals (steel 45) in the vacuum 10^{-3} Pa (2) and in the air (1). The extent of the electrification in the vacuum is higher than in the air, and the triboelectrification current as a function of time has a clearly pronounced impulse nature.

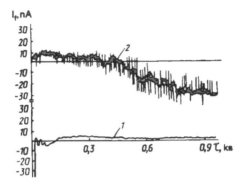

Fig. 6.8. Kinetics of electrification current during friction of high density polyethylene in air (1) and in vacuum $P_v = 1.3 \cdot 10^{-3}$ Pa (2)

It has been confirmed that the positive charge becomes negative already in the "moderate" vacuum and lowered humidity [23]. Figure 6.9 shows the curves of triboelectrification current variations when the reversed PCA-steel couple operates in the vacuum with various degrees of rarefaction. It is apparent that the general pattern of current variations during friction in the vacuum and in the air are identical, since initially the electrification current increases, and afterwards it reduces and undergoes inversion many times, just like in the air.

The analysis of the current diagrams indicates that in the general case the degree of electrification of polymers goes up with the increasing degree of rarefaction ($I_f = 40$ nA at $P_v = 7.6 \cdot 10^4$ Pa; $I_f = 110$ nA at $P_v \cdot 10^{-3}$ Pa). Typical current variations at $P_v = 1.3$ Pa are apparently due to the effect of the adsorbed water layers once their thickness exceeds the size of molecules. Moreover, the nature of electrification current during friction in the medium and high vacuum in the initial period (before inversion) has a "pulsing" nature due to the emission of high energy electrons when the metal-polymer contact is disrupted.

Investigation of the contact mode during the friction of the metal-polymer couples in the vacuum (Fig. 6.10) indicates that the growing load leads to a greater absolute maximum of the electrification current which reduces since the moment friction starts until the first inversion. Yet, the

158 ENVIRONMENTAL EFFECTS

current inversion during friction in the vacuum occurs at the surface temperature of contacting bodies being 323...333 K. This temperature range corresponds to the PCA glass transition range in the air of 350-380 K [24-26] (T=328 K). So, this can serve as a proof that significant modifications occur at the moment of inversion both in the surface layers, such as water desorption, and in the volume of the polymer.

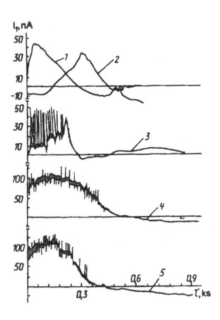

Fig. 6.9. Electrification current variations in response to time of operation of the reversed polycaproamide-metal friction couple at rarefaction P_V=10^5 Pa (1); 13.0 (2); 1.3 (3); 1.3 10^{-1} (4) and 1.3 10^{-3} Pa (5)

The triboelectrification current inversion relates to the contact mode and environment, and it can be caused by the change of the dominating type of charge carriers due to structural and chemophysical transformations in polymers. The phenomenon of the current inversion during the friction of dielectrics is explained by the change of the charging mechanism when the temperature is increased. It has already been demonstrated that the first electrification current inversion is caused by moisture desorption from the surface layers of a polymer.

A mass-spectrometric analysis has been performed for the direct experimental proof of the hypothesis of the frictional interaction between polymers (PCA, PTFE), on metal (Steel 45) in the vacuum 10^{-3} Pa. During friction of PCA in the reversed couple with the metal an intense liberation of volatile substances is observed with their mass-spectrometric composition [25] close to the composition of the volatiles of the thermodestruction polymeric products.

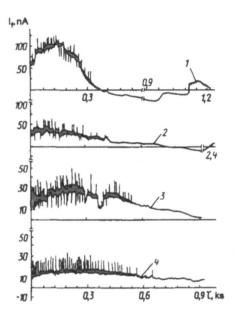

Fig. 6.10. Kinetics of electrification current during friction of polycaproamide-metal reversed couple at sliding velocity v = 0.3 m/s and nominal pressure p equal to 0.5 MPa (1); 0.375 (2); 0.25 (3); 0.125 (4)

. The major portion of the volatiles is water (m/e=18) from the original polymer. The curve of water liberation as a function of temperature in the friction zone (Fig. 6.11) shows a peak in the region of the first triboelectrification current inversion.

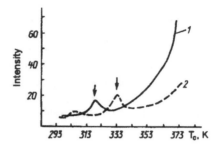

Fig. 6.11. Intensity of water liberation in response to temperature of contacting bodies during friction of polycaproamide (*1*) and polyethylene terephtalate (*2*) in vacuum P_a = 1.3 10^{-3} Pa (arrow shows moment of triboelectrification current inversions)

Thus, the study of the electrophysical phenomena during friction of polymers in the vacuum has covered the effect of rarefaction pressure upon triboelectrification of polymers. During the friction in the vacuum the electrification current magnitude and sign depend upon the pressure in the vacuum chamber. Load increase leads to the greater current and shorter times until the first inversion. The kinetics of the electrification current during the friction of polymers in the air and in the vacuum have similar nature, and the first electrification current inversion is caused by water desorption from the surface layers of a polymer.

REFERENCES

1. A.I. Gubkin. *Electrets* (in Russian), Moscow, 1978.
2. V.N. Klassov, and K. A. Osipov.*Unsteady Processes in Dielectric Materials* (in Russian), pp. 93-98, Moscow, 1983.
3. J. Turnhout. *Thermally Stimulated Discharge of Polymers Electrets*. N.Y.,1975.
4. G. Cessler.*Electrets* (English translation), pp. 8-104, Moscow, 1983.
5. Yu.A. Khrustalev. *Unsteady Processes in Dielectric Materials* (in Russian), pp. 47-53, Moscow, 1983.
6. L. Loeb. *Static Electrification* (in Russian). Moscow-Leningrad, 1963.
7. J. Lowell, and A.C. Rose-Innes. *Advances in Physics*, vol. 29, no 6, pp. 947-1023, 1980.
8. O.N. Sheverdyaev. *Antistatic Polymeric Materials* (in Russian), 176 pp., Moscow, 1983.
9. Yu.I. Vasilenok. *Protection of Polymers Against Static Electricity* (in Russian), Leningrad, 1981.

10. V.E. Shashoua. *J.Polymer Sci. Part A*, vol. 1, no 1, pp. 169-187, 1963.

11. M. Beyer, D.I. Eckhardt, and Q. Lei. *Etz. Arch.*, vol. 7, no 2, pp.40-49, 1985.

12. Q. Lei. *Conf. Rec. Int. Conf. Prop. and Appl. Dielect. Mater.*, vol. 2, pp. 421-424, 1985.

13. M. Ononda, H. Makagama, and K. Amakwa. *Trans. Inst. Elec. Eng.Japan*, A108, no 5, pp. 225-232, 1986.

14. J. Lopes, B. Despax, and G. Mayoux. *Proc.Ind. Int. Conf. Conduct and Breakdown Solid Dielect.*Erlayner. N.Y., pp. 191-195, 1986.

15. O.A. Myazdrikov, and V.E. Manojlov. *Electrets* (in Russian). Moscow, 1962.

16. S.I. Guzenkov. *Soviet Journal of Friction and Wear*, vol 11, no.1, pp. 151--154, 1990.

17. A.F. Klimovich, and V.S. Mironov. *Soviet Journal of Friction and Wear*, vol 6, no. 5, pp. 18--26, no 6, pp. 52--58, 1985.

18. V.A. Belyi, L.S. Pinchuk, A.F. Klimovich, and S.S. Guzenkov. *Proc. 5th Int. Congress on Tribology*, pp. 276-281, Helsinki, 1989.

19. A.I. Gubkin. in 'Radioelectronic Materials' (in Russian), Moscow, 1986, p. 45.

20. Summ, B.D. and Gorjuonov, Yu.V. 'Chemophysical Principles of Wettening and Spreading' (in Russian) , Moscow, 1979.

21. B.B. Damaskin, S.A. Ptresey, and V.V. Batrakov. *Adsorption of Organic Compounds on Electrets* (in Russian), Moscow, 1968.

22. V.N. Anisimova, T.N. Vladykina, B.V. Deryagin, and Yu.P. Toporov. *Proc. of the VIII All-Union Symp. on Mechanical Emission and Mechanochemistry of Solids*, pp. 173-178, Tallinn, 1986.

23. P.A. Thiessen, and K. Sieber. *Phys. Chem.*, vol. 260, p. 410, 1979.

24. A.F. Klimovich, and V.S. MIronov. *Soviet Journal of Friction and Wear*, vol. 2, no 3, pp. 128--131, 1981.

25. A.F. Klimovich, and S.N. Guzenkov. *Soviet Journal of Friction and Wear*, vol. 10, no 5, pp. 6--11, 1989.

26. Guzenkov, S.N. and Klimovich, A.F. *Journal of Friction and Wear*, 1992, 13, no.3, 52–56.

CHAPTER 7

APPLICATION OF THE TRIBOELECTRIFICATION AND ELECTRET EFFECTS IN POLYMERS

Recently, triboelectrification and electret effects in polymers have been taken into account and began being used more and more in practice. Traditionally, there were two trends – to use positive effects of these phenomena and to fight against their negative effects. Initially, the trends were developed relating to the electrization phenomena and only as of the 1980's have the electret effects drawn any attention within the scientific community.

7.1. FRICTION ASSEMBLIES

Electric phenomena are the most widely used to control friction parameters at contacts of polymers and metals. It is common to suppress triboelectricity by combining the polymers of different polarities or by passing the current, rising the polymers conductivity or setting the preliminary electret state. Triboelectric processes play an important part in friction and selective transfer, as well as in frictional hydrogenization.

Bilik [1] has reported interesting results on the effect of electrostatic charging on wear of the metal-polymer friction pairs (schemes I–IV in Fig. 2.1). The most intensive wear of the counterbody occurs at closed circuit or when the external voltage is applied in series. In the latter case the charge transfer is more intensive. Applying the opposite voltage and suppressing this process, the lowest wear rate is achieved (Fig. 7.1 b, curve 6).

Friction in direct and reverse schemes [wood-polymer composite (WPC) with 80% of graphite against steel] decreases when electric current passes through the pair. Minimum friction for the reverse scheme is

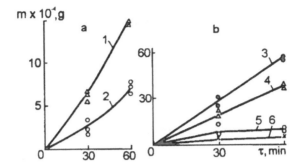

Figure 7.1. Wear of metal bearings in friction pairs wood-polymer composite (with graphite) – steel (a), Epoxy – steel. 1, the pair is closed as in scheme II, Fig. 2.1; 2, the pair is disconnected as in scheme II, Fig. 2.1; 3, the pair is closed as in scheme III, Fig. 2.1; 4, the pair is closed as in scheme II, Fig. 2.1; 5, the pair is disconnected as in scheme I, Fig. 2.1; 6, the pair is closed as in scheme IV, Fig. 2.1.

achieved at 3 A. Higher current does not decrease the minimum because momentary charge-discharge pulses (MCDP) are fully suppressed on the friction pair nominal contact.

According to the data from [2], the friction in pair "fabric laminate – steel" was reduced by 40% on average and the temperature decreased after the discharger application.

One of the ways to suppress the static electricity is to combine the electropositive (E^+) and the electronegative (E^-) polymers in parts production. This results in the total charge reduction. For example, if the shaft is made of composite containing 60–70% of polymetylmethacrylate (PMMA) (electronegative polymer) and 40–30% of polytetrafluorethylene (PTFE) (electropositive polymer), the lowest level of electrization is achieved.

When the amount of electronegative and electropositive polymers in the substance is approximately equal, friction decreases to its optimum value.

Nevertheless, no combinations of polymers with different polarity can give zero electrization. To suppress MCDP, composites with electroconductive fillers (graphite, soot, metal powders) were used and the electric current was passed through the contact zone . It was mentioned that the triboelectricity suppression may decrease friction in the pairs by up to a factor of 10. When the electropositive polymers are used, they "smear" over the metals. If a polymer is electronegative, it is metallized. Combining the electronegative and the electropositive polymers is accompanied by reduction of the electrization level friction, and the temperature of the composites. With increase of the polymer conductivity (for example, by introduction of graphite, soot, metal powders, and so forth) the magnitude of total potential decreases. For instance, if 100% epoxy resin and 50% graphite are added, the potential reduces from 2000 V to 500-400 V and when 100% of graphite is added, the potential reduces to 50-100V.

There is a relation between variations of the surface conductivity and the wear of polymer materials. Decreasing ability of polymers to generate and accumulate static charge at the growth of their conductivity is usually attributed to the smaller number and the smaller area of contact spots at the contact with a metal (spatial structures – conductive chains of the filler are formed in the polymer). The three component scheme (metal – filler – polymer) is formed characterized by the increased quantization of the polymer-metal contact. Total potential also decreases due to leaks created by the increased conductivity. Higher conductivity results in smaller values of the potential, but the higher values of the current result from one body rubbing another. A lubricant also sharply lowers the level of the rubbing bodies electrization and does it faster when its own conductivity is higher.

Changes in the state and the properties of the polymer surface layer, particularly, variations in thermal and electric conductivities have been reported [3,4]. The authors have attributed these processes to the alteration of electric state in the polymer volume, in particular, to the growth of the number of charge carriers [5].

Thus, the friction of metal-polymer pairs is accompanied by significant electrization, the rate of which is defined by chemical and physical nature of the solids. Suppression of the charges generated is accompanied by an 80-90% decrease of the friction force and the temperature reduction of the rubbing materials. Their wear depends on the direction of the electron motion in the contact zone: the material losing electrons is worn more intensively.

Friction of polymers on metals usually results in formation of polymer film on the metal. It has been shown that wear of the metal and the polymer film after the frictional transfer optimization becomes minimum. This is used for designing of self-lubricating polymer composites [6–10]. The mechanism of the transfer of polymer onto metal and metal onto polymer is very complicated and is governed by mechano- and thermodestruction, fatigue phenomena and the polarity of the electric charges generated [6–13].

According to [3], sign-alternating transfer originating from the electric nature of the mass-transfer in the "polymer – metal" system takes place due to the ability of the metal to control this process depending on the "electropositive" or "electronegative" properties of the polymer. As a result, products of mechanodestruction of the electropositive polymer in the friction zone are transferred first in one direction, and then in the other until continuous films formed on the mating surfaces. For example, PTFE with an electrization potential of +500 V, used as a source of solid lubricant, allowed a hardly noticeable wear of the steel – ebonite pair (electrization potential –600 V) under conditions of dry friction.

Interesting results have been obtained when the products of thermal-oxidative destruction of polyethylene (PE) (pyrolysis at 400 °C) were used as an interlayer in the metal-polymer friction pair [11]. Their introduction into the friction zone of the reverse "fabric laminate – steel" pair (sliding velocity 1 m/s, load 0.65 MPa) led to sharp drop of the friction coefficient to 0.05 and lowering of the triboelectric charge.

The influence of polymer additives [polyisobutylene (PIE), PE, PTFE] to Vaseline oil on friction of steel – polycaproamide (PCA) pair (sliding velocity 0.5 m/s, load up to 40 MPa) has also been studied. The results have shown that the electronegative additives of PTFE and PE were adsorbed on the electropositive PCA and reduced its triboelectric potential. On the contrary, polyisobutylene additive has increased triboelectric potential on the PCA surface.

With respect to minibearing separators of the PTFE-based composite, it has been shown [14] that triboelectrification, on the one hand, retains wear debris in the friction zone and favors the formation of the protective films, and, on the other hand, may lead to friction increase and jamming.

The negative effect of the static electrization in the metal-polymer friction assemblies is often connected to hydrogen wear. It is known that the detachment of hydrogen at polymer friction results in hydrogenation of the steel surface and makes it brittle [15,16]. The triboelectrification promotes the penetration of protons with very small size ($1 \cdot 10^{-13}$ cm) into the metal lattice having an opposite charge. There is a relation between the charge sign on the worn surface of the polymer and the hydrogenation. Positive charging corresponds to a two-fold to three-fold increase in the hydrogen content [12].

Studies of the dependence of the metal surface wear on hydrogene detachment, which increases when the decomposition of the lubricant and the water molecules sets in during friction, have afforded the following recommendations. The use of additives whose molecules contain dipole groups ending with hydrogen in lubricants should be restricted. Rubbing surface should have positive potential to repel H^+, in this case, its neutralization will take place in the lubricant volume.

The process of hydrogen wear is closely tied with the "wearless effect", called selective transfer [15]. This phenomenon consists in formation of the copper film of 1 to 2 µm thick on the rubbing surfaces of a steel – bronze friction pair lubricated by special substance. It accompanied by formation of micells in the lubricant. Those are charged copper particles that are precipitated on the surfaces in the friction zone forming servovite film*. In the steady-state mode, the film fragments migrate from one surface to another. Wear products are retained in the gap by electric forces. Servovite film can also be formed at the friction of polymer composite materials, for instance, PTFE filled with copper protoxide or composition on the base of epoxyfuran oligomers and copper-containing fillers.

Polymerization at the contact is one of the characteristics of selective transfer. Properties of the polymerized layer are determined by with the servovite film mobility. It conditions the preservation of pseudohydrodynamic effect at friction path and the distribution of contact pressure over a larger area in comparison with the friction of metal against polymer without such mobility. In its turn, the mobility of servovite film affects friction coefficient which depends on the internal friction inside the film. Moreover, electric charge of particles and mating surfaces plays a certain role, especially at initial stages of friction.

Adding to the list of negative effects of static charge accumulation in tribosystems, it should be mentioned that the charged polymer surfaces are capable of attracting dust, moisture, and wear debris from environment. Discharges when approaching the grounded metal parts, electric aging of polymers and electric erosion of metals are also possible. To prevent an accumulation of electric charges, it is necessary to reduce the polymer resistance to 10^6 Ohm·cm or lower. This can be achieved by using conductive polymer composites and lubricants [10,18–24]. Nevertheless, the electric discharges between friction pairs conditioned by the friction nature cannot be removed completely.

---

* It has been shown in work [17] that this process is not universal.

We should mention the possibility of controlling electric phenomena in sliding electric contact pairs of composites, for example, in variable wireless resistors (VWR). One of parameters defining the resistors adjustment is the motion noise level (MN) that appears at frictional interaction of contact spring against resistive element (RE). Studies have shown that the composite REs with nominal resistance above 1 MOhm can acquire an electret state. Magnitude and polarity of the electrization current depend on the RE composition and properties. High-resistance REs exhibit correlation between the electrization current magnitude, the resistance, and MN level. The MN level can be decreased by applying liquid lubricant with specific volume resistance not higher than 10^2 to 10^6 Ohm×m.

One of the latest applications of the electrization is controlling the frictional parameters of polymer material [25]. It is realized due to electrostatic interaction between volume charge of the electret polymer and the volume charge induced by friction. Depending on the signs of the electret charge and the tribocharge, the friction coefficient and wear rate of the polymer material may be lower or higher in comparison with its initial state.

Table 7.1. shows the friction force F and the mass wear rate J_g vs. the magnitude and the sign of effective surface charge density (ESCD) for thermoelectret coatings.

Table 7.1. Friction force F and wear rate J_g of the thermoelectret coatings.

Coating material					
High density polyethylene			Polycaproamide		
Effective surface charge density, μKl/m^2	F, N	J_g, mg/(m^2·m)	Effective surface charge density, μKl/m^2	F, N	J_g, mg/(m^2·m)
0	3.5	1.35	0	2.9	3.87
+10	3.3	0.78	+1.6	2.7	2.32
+20	3.0	0.68	+6.0	2.1	2.70
+40	3.1	0.55	+13	2.1	3.47
+60	3.3	1.08	+20	3.0	–
+10	3.6	1.01	–1	2.9	1.88
–20	3.5	0.87	–3	2.8	1.86
–40	3.6	0.49	–8	2.2	–
–60	3.8	1.01	–20	2.8	–

Depending on the magnitude and the sign of the electret charge, friction in the coatings and their wear rate may be decreased by 10–30% and by 1.5–3 times, correspondingly. Preliminary electretization may be promising in modeling the electric state of the frictional contact and in clarifying its part in the polymer-metal pairs functioning.

There are some effective means to control the triboelectric characteristics decreasing friction and wear. These include the following:
1. Suppression of triboelectricity by passing of the electric current through the friction zone.
2. Decrease of the triboelectrification by selection of the optimal electropositive, electronegative and electroconductive components for tribocomposites.
3. Preliminary electretization of the polymer surfaces.
4. Application of electroconductive lubricants and surfactants.

7.2. DIAMOND CUTTING

Polymer electret materials appeared to be effective in diamond cutting [25–27]. It is known that the diamond is a dielectric with specific resistance of 10^{12} to 10^{14} Ohm×m and dielectric constant $\varepsilon = 5.7$. So, during friction, diamonds acquire large electrostatic charge. The diamond has unique properties but is very difficult to process.

Diamond crystals are cut up with special machines by thin tin-phosphorus bronze disks charged with diamond powder [28]. Destruction of the diamond at cutting is due to its dynamic contacting with the edge of the disk. Speed at the disk edge reaches 35 to 55 m/s (disk of diameter 65–76 mm revolves at 10,000–14,000 rpm). Such contacting is characterized by intensive electrization and accompanying electrophysical phenomena that have a significant effect on the cutting process.

Studies of these phenomena [29] have revealed the presence of an electric current of several hundred milliamperes in the diamond-instrument-ground circuit. Its magnitude and direction depend on physico-mechanical properties of the abrasive layer binder. As binders, polymers with different electrophysical properties, polymer composites, mineral oil, copper-zinc alloy, and nickel galvanic platings are commonly used. Test results for different binders are presented in Table 7.2.

Electrization current at diamond cutting by the disks without coating and disks with binders based on mineral (castor) oil reaches 10 to 30 nA.

Table 7.2. Electrization current at diamond cutting by disks with different coatings.

Disk diamond coating binder	Polarity	Magnitude, nA
Disk without coating	+	10-30
Mineral oil	+	10-20
Copper-zinc alloy plating	+	40-60
Nickel plating	+	100-120
PTFE (Ftoroplast – 4)	+	30-40
Polyacrylonitrile	+	100-130
Polystyrene	+	200-300
Lacquer PAK–1M	+	200-300
Phenol formaldehyde resin	+	300-500
Lacquer PAK–1M modified by graphite and molybdenum disulfide	+	200-350

Its magnitude rises by 2 to 3 times at the presence of nickel plating and by 5 to 6 times for copper-zinc one. Polymer binders significantly increase the electrization current, and PTFE binder alters the current polarity from positive to negative. Binders based on polyimide (PA) lacquer, polystyrene (PS) and phenol formaldehyde resin (PFR) increase the electrization current to 200 to 500 nA and higher. It should be mentioned that the potential on the mandrel measured by electrostatic voltmeter S–50 reached 200 V.

Electrization current kinetics at diamond cutting by the disks without coating and charged with binders based on castor oil and polyimide lacquer are shown in Figure 7.2. Irrespective of the binder type, electrization current curves have an extreme character governed by the diamond cutting process, namely, the gradual cutting into the diamond crystal, increase of the contact area, and its successive reduction.

Experiments have revealed that complete insulation of the instrument and the diamond results in sharp reduction of the electrization current, in cutting speed decrease by 18 to 20% and approximately in twofold increase of the disk durability. Analysis has shown that highly productive instrument should have an easily-electrifying and wear-resistant binder, whereas the diamond fixing elements and the instrument should be grounded.

The main parameters of the diamond cutting for different types of the disk coating binders are represented in Table 7.3. Studies have shown that the electrization intensity (i.e. the electrization current) correlates with the

APPLICATION OF THE EFFECTS

cutting rate, the cutting disks durability, and the quality of the products [27].

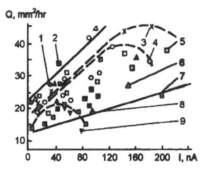

Figure 7.2. Cutting rate vs. electrization current for the different disk coating binders: 1, castor oil; 2, polytetrafluorethylene; 3, polyimide lacquer + graphite + MoS_2; 4, polyimide lacquer; 5, phenol formaldehyde resin; 6, polyimide lacquer + phenol formaldehyde resin; 7, polystyrene; 8, polyimide lacquer + polystyrene; 9, polyacrylonitrile.

Table 7.3. Characteristics of the diamond cutting process.

Disk coating binder	Electrization current I, nA	Cutting rate Q, mm^2/hr	Raw material losses, %	Processing quality
Castor oil	+30	18.14	9	Good
Polyimide lacquer	+280	20.18	7	High
Polyimide lacquer + graphite + MoS_2	+320	21.05	6	Very high

Experimental cutting rate Q was determined as

$$Q = S/t, \quad (7.1)$$

where S is an area of the cut up surface, mm^2; t is cutting time, hr.

The tests have shown (Fig. 7.2) that in general, the higher cutting rate corresponds to the higher electrization current. Experiments on cutting up 65 diamonds (5 to 10 crystals for each type of the disk coating) under different conditions have revealed the increase of the cutting rate with the electrization current irrespective of disk coating paste type. However, the character of the approximation curve for the dependency $Q = f(I)$ remained unclear. Other experimental series where 162 crystals were cut up have shown that the dependence is of a parabolic type for all the cases. For example, for disks charged with binder based on castor oil, the dependency of the cutting rate on the electrization current can be approximated by square or cubic parabola (Fig. 7.3):

$$Q = 23.31 + 0.22\, I - 1.87 \cdot 10^{-3}\, I^2 + 3.46 \cdot 10^{-6}\, I^3. \qquad (7.2)$$

Within certain approximation, the dependence $Q = f(I)$ for polymer based binders can also be rendered as a parabolic (dashed lines in Fig. 7.2). Hence, the relation between cutting rate and electrization current has an extreme character governed by the effect of the current on the cutting instrument.

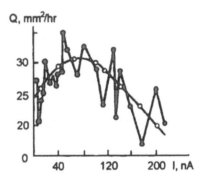

Figure 7.3. Cutting rate vs. electrization current for disks with the coating binder based on castor oil. Parabolic curve is an approximation of the experimental data.

Crystal destruction occurs due to the abrasive mechanical action, diffusion (thermal-oxidative) and adhesive wear. It is known that the local microcapacitor fields forming in the contact zone under dynamic interaction of solids activate diffusion and frictional transfer and determine the destruction process. In addition, the electric field intensifies the destruction of the polymer macromolecules and increase polymer adsorbability. Taking into account the electromechanical (electroerosion) wear theory of solids [30,31], the increase of the cutting rate with the electrization current can be explained by the activation effect on the processes of adsorbtion-induced strength reduction of the diamonds at contact with a polymer, diffusion and adhesive wear and electroerosive destruction of the crystals.

However, positive effect of electrization current on the cutting of diamonds is accompanied by its negative influence on the instrument durability. For instance, at electrization current of 0.5 (for insulated instrument), 25 and 80 nA, the average durability of the disks charged with pastes based on castor oil becomes 305, 190, and 153 mm^2, correspondingly. As a result, high electrization current leads to lowering the cutting rate due to worsening cutting properties of the disk.

Figure 7.4 represents cutting rate vs. electrization current for instruments charged with diamond pastes with conventional (castor oil) and composite (PAK-1M added by PFR, graphite and molybdenum disulfide) binders. The results have been obtained under equal conditions for identical diamonds.

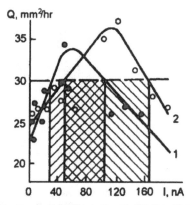

Figure 7.4. Cutting rate vs. electrization current for the disks charged by the diamond pastes with conventional (1) and composite (2) binders. Shaded areas correspond to the optimum cutting rate.

In particular, specific load was carefully controlled and kept constant because the cutting rate is defined primarily by this parameter. Considering the optimum cutting rate of $Q = 30$ mm^2/hr, the optimum range of the electrization current can be derived. As it can be seen from Figure 7.4, the electrization current working ranges are 45 to 165 nA and 30 to 105 nA for composite and castor oil binders, correspondingly. Beyond these ranges, the cutting rate is much lower. Higher values of Q reaching 37 mm^2/hr are characteristic for the instrument charged by the paste with composite binder. The same conclusion follows from Figure 7.2, where maximum cutting rate for the instrument is above 45 mm^2/hr. Moreover, the character of the descending branches of the curves in Figure 7.4 indicates higher electroerosive durability of the cutting disks charged by the pastes with composite binders.

Studies have shown that cutting disks coatings with composite polymer binders are promising. They have high cutting rate, wide electrization current working range, and high electroerosive durability.

The method of mechanically cutting diamonds in the presence of the electric field generated by electret linings should also be mentioned. Necessary equipment consists of the cutting plate, supporting elements on the plate sides and at least one electret polymer lining sheet between each of the elements and the plate. The electret has ESCD of 10^{-6} to 10^{-4} Kl/m^2 and dielectric permeability of 2.1 to 7.5 with a sheet thickness being 0.01 to 1 mm. Mating electret sheets should be placed with the opposite charge signs inside.

Experiments on diamond cutting with and without the electret linings (the crystals were identical on their average weight, sizes and quality) were conducted at constant load on the disk of 1.2 N and speed of 13,000 rpm. Cutting time t, cutting area S, and the visual quality of the cutting were estimated. Results obtained are presented in Tables 7.4 and 7.5.

Table 7.4. Diamond cutting rate for instruments with and without the electret linings.

Disk coating binder	Instrument	Diamond mass, carats	Cutting rate, mm^2/hr	Cutting quality
Castor oil	With electret linings	0.46–0.53	19.646	High
		0.25–0.30	19.068	
	No linings	0.46–0.53	18.191	Good
		0.25–0.30	17.344	
Polyimide	With electret linings	0.46–0.53	21.242	Very high
		0.25–0.30	22.182	
	No linings	0.46–0.53	19.148	High
		0.25–0.30	19.066	

Thus, electret linings provided higher (by 9%) cutting rate, lower (by 5 dB) noise level, and reduction of the raw material losses by 0.1%. Better quality of cutting allows reduction of raw material losses at successive operations, particularly at finish grinding, and also the grinding time itself. Also, in this case it is not necessary to use special electric system, and the instrument is more vibration-proof.

Table 7.5. Dependence of the cutting parameters on the electrophysical characteristics.

Average effective surface charge density (of the electret lining, Kl/cm²	Electric field intensity between the plates, kV/cm	Cutting rate, mm²/hr	Cutting quality
$4 \cdot 10^{-4}$	400	22.330	Very high
$1 \cdot 10^{-4}$	100	22.218	Very high
$6 \cdot 10^{-5}$	60	21.984	Very high
$1 \cdot 10^{-5}$	10	21.013	Very high
$3 \cdot 10^{-6}$	3	19.215	High
Without linings	0	19.153	High

7.3. POLYMER COATINGS

The established laws of dispersed   dynamic electrization of polymers allowed design different techniques and devices for polymer coatings deposition. Basic elements of these techniques are tribocharging devices [32,33].

A major trend in tribocharging pulverizer development is an improvement of the polymer particles charging efficiency. It is reached by intensification of the particles' contact interaction with the charging element surfaces. Different nozzles and transporting pipelines as well as special devices are used for these purposes.

The first approach is characterized by the   continuous electrodeposition. Passing through the pipelines, the particles are charged and then are deposited on the grounded article with the help of a pulverizer. Contact interaction may be intensified in this scheme by rotating different parts of the pipeline or by increasing turbulence of the air-powder flow in the   widened part of the pipeline and also by the flow rotation.

Figure 7.5 represents a device in which particles are charged in helical grooves formed by two concentric dielectric tubes one of which is rotating. Charging is intensified due to sliding and diffusion discharges from the inner tube surface.

High charges accumulating on surfaces of the dielectric charging elements discharge periodically. It defines instability of the dispersed polymer charging. To avoid this, the charging element outside is coated with conductive material and is grounded.

At long operation, walls of the charging elements can undergo local destruction because of sparkthrough or sliding discharges. Holes in the walls filled with conductive material and connected to the external grounded layer can help to prevent this effect.

The second approach is based on the use of additional devices and is characterized by periodic operation. Air-powder flow circulates inside the closed device until necessary charge of the particles is reached. Then the device is connected to the tanks with the grounded articles inside or directly to the pulvirizer.

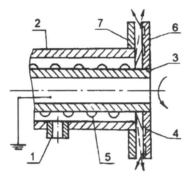

Figure 7.5. Tribocharging device. 1, connecting pipe for air-powder pulp; 2, external stationary tube; 3, inner rotating tube; 4, 5, helical grooves; 6, 7, flanges.

Significant charging of the device metal walls due to continuous circulation of the air-powder flow is sometimes used to create high-voltage potential in pulvirizer or on electrodes of pseudoliquefying chambers.

To ensure stable charging of dispersed polymers (independently of the pulvirizer type and kind of the charged particles), the exact dosage of the powder received by the pulvirizer is necessary. It is especially important for automatic modes. Dosage devices must provide stable feeding of pulvirizer with the powdered polymer and wide-range control for the polymer consumption and the flow speed. Ejection systems used in electrostream processes meet these requirements.

The electrostatic apparatus in Figure 7.6 consists of electrostatic generator and tank (7) for the coating deposition. The generator with closed circulation of air-powder pulp consists of an electric motor (10), chamber (1) with activating fan (2), circulation pipes (3) and collector (4). High voltage is transmitted from the collector to charging device (6) rotating with the help of the motor and the flexible shaft (9). Item (8) must be grounded. The high-speed circulation of the particles provides the generation of high voltage (up to 100 kV and above). Such an apparatus demands no compressed gas or special sources of high voltage.

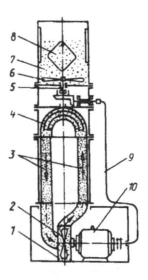

Figure 7.6. Scheme of electrostatic apparatus for polymer coating deposition. (1) chamber; (2) activating fan; (3) circulation pipes; (4) collector; (5) barrier; (6) charging device; (7) tank; (8) item; (9) flexible shaft; (10) an electric motor.

In the hand-held devices for polymer coating deposition, the polymer powder is electrified by impact and dynamic interaction of the powder particles with the surface of the charging unit. The charging unit is the main element of such devices. The most common design of the hand-held device is presented in Figure 7.7. It contains case (1) with the channels for gas-polymer pulp feeding, handle (2), coaxial rotor electrodes (3) and (4) with blades (5), guiding nozzle (6), mixer (7) with the outlet duct (8) connected by hoses (9) to gas reservoir (10) and pseudoliquefying chamber (11). External electrode (3) is fixed in bearing (14) while the inner electrode revolves on axis (15). Device also contains tap (12) on the air hose and valve (13).

Dispersed polymer pseudoliquefied by compressed gas in chamber (11) moves to mixer (7) and then through the outlet duct into the working area. Additional gas feeding from reservoir (10) is controlled by tap (12). Gas-polymer dispersion under high pressure moves at high speed onto the rotor electrodes. Due to multiple contacts of the polymer particles with metal electrodes, the charges are generated on the particles. To achieve high charge on the polymer particles, the speed of gas-polymer dispersion at the electrode intereface inlet should be above 10 m/s.

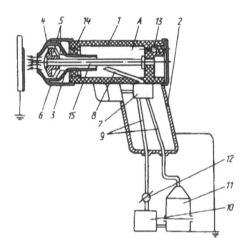

Figure 7.7. Scheme of the hand-held device for deposition of polymer coatings. (1) case with the channels for gas-polymer pulp feeding; (2) handle; (3, 4) coaxial rotor electrodes; (5) blades; (6) guiding nozzle; (7) mixer; (8) outlet duct; (9) hoses; (10) gas reservoir; (11) chamber; (12) tap; (13) valve; (14) bearing; (15) axis.

Such a device allows the high-output deposition of the coatings on cold surfaces with successive melting. Complex articles and the whole assembled unit can be coated as well. The process is easy for automation..

To improve the device design and the coating quality, the electrode working surfaces or whole rotors can be made of electret materials. Then these elements will generate external electrostatic field.

Thin (2–20 nm) polymer coatings can be deposited in vacuum with an additional processing in corona discharge [34]. The resulting coating has an electret charge with the density of 4 Kl/m^2. Such a characteristic is sufficient for the use in electronic devices. The electret state, as a rule, is accompanied by specific piezo- and pyroelectric properties that enable the use of the thin-film electret as a sensor.

7.4. ELECTRET FILTERS

Electrization and electretization of melted polymers during their treatment can be applied in the manufacture of different articles, particularly, polymer fiber filters. Pneumoextrusion is one of the effective methods for obtaining the polymer fiber filters [35]. Thermoplastic polymer melted in the extruder to fluid state is moved to a spraying device. During this operation the fiber formation and the drawing take place. The polymer undergoes the triple contact interaction: when it moves inside the extruder, during pneumodrawing, and when it interacts with the filter material. This results in the formation of the electret state in filters. The magnitude and stability of the electret charges may be significantly affected by additional discharge at the extruder outlet[*].

All spectra presented in Figure 7.8 and Figure 7.9 show the maximum current in the range of 373 to 383 K. Positive polarity of these peaks is characteristic for spontaneous electrization at extrusion and for electrization by friction (Fig. 7.8, curves 1–3) as well as for electrization by positive corona discharge at extrusion (Fig. 7.8, curve 4 and Fig. 7.9, curve 2). Electrization in negative corona discharge at the extrusion alters the peak polarity to negative (Fig. 7.8, curve 5 and Fig. 7.9, curve 3).

These data show that the principle of electret-triboelectrification superposition first revealed for friction of polymers is also applicable to the electrization at extrusion. Similar results were obtained at extrusion of PCA filters (Fig. 7.10). Simple preliminary treatment of the raw material granules increases the amount of structural defects, i.e. the concentration of volume and surface electron states. Their ionization at the extrusion leads to formation of the stronger electret charge (Fig. 7.11). The charge can significantly drop (by 20–40%) during the first two days after the exposition, and afterwards it becomes steady.

Experiments with the air filters made of electret materials obtained according to the above method have shown higher (by 10–15%) effectiveness of air straining in comparison with conventional porous materials with the same physico-mechanical characteristics. By choosing the optimal materials and filter design, the higher efficiency can be achieved [23].

---

[*] The experiments were performed under the guidance of V.P.Shustov.

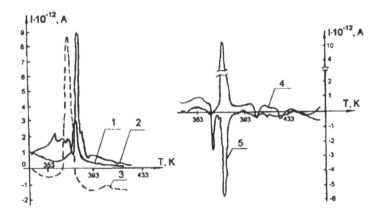

Figure 7.8. Thermostimulated current diagrams for polyethylene films electrified by friction with wool (1), in positive corona discharge (+6 kV/cm) (2) and for fiber filter electrified at pneumoextrusion: spontaneously (3), in positive (4) and negative (5) corona discharge. All measurements were done a day after the materials were manufactured.

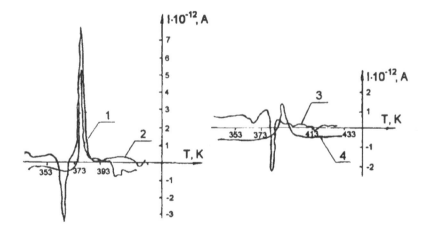

Figure 7.9. Thermostimulated current diagrams for polyethylene filters electrified at extrusion: spontaneously (1, 4), in positive (2) and negative (3) corona discharge. Measurements were done two days (1-3) and six days (4) after the materials were manufactured.

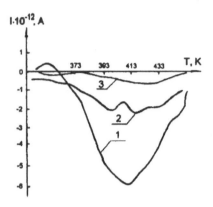

Figure 7.10. Thermostimulated current diagrams for polycaproamide filters. The material was electrified: spontaneously at extrusion (1); in negative corona discharge at extrusion (2); in positive corona discharge after extrusion (3).

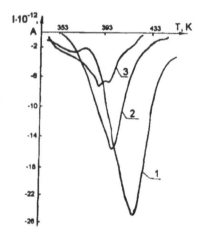

Figure 7.11. Thermostimulated current diagrams for polycaproamide filters (raw material was dried at 373 K). The material was electrified: spontaneously at pneumoextrusion (1); in negative (2) and positive corona discharge after extrusion (3).

Such filters have 2 to 4 times longer service life than conventional ones. In addition, they are sealed much later, their dust capacity is 1.5 to 2 times higher and filtration rate is up to two times greater.

Fiber electret filters manufactured by pneumoextrusion and filled with different dispersed particles are also highly effective. Particularly, fine-dispersed ferrite fillers [36,37] can add the magnetic mechanism to the particle filtration process (Table 7.6). Magnetic-electret fiber polymer materials have much better functional characteristics than the conventional ones used in automotive, machine tool, and other industries.

Table 7.6. Percentage of the particles in oil after filtration.

Filter material	Particle size, μm				
	0–5	6–10	11–20	21–30	31–40
Magnetic fiber	100.0	–	–	–	–
Corrugate paper	96.53	2.60	0.43	0.22	0.022

7.5. POLYMER COMPOSITES

Dispersed polymers can be effectively used as matrix material in the manufacturing of highly-durable fiber polymer materials [32].

Traditional technology in the manufacturing of such materials consists in combining discrete polymer matrix with fillers by scattering and spreading of pastes in water environment.

A proposed electron-ion technology (EIT) is based on the electrostatic combination of the components. This process has the following stages: the adhesive contact of the binder particles with fiber filler, the particles electrodeposition in a layer with high redundant charge and, finally, the formation of the continuous polymer matrix . The latter is realized in the electric field of the redundant charge of the binder at thermal treatment and is accompanied by transition of the binder and the dielectric filler into the electret state.

Triboelectrification is often used for charging the powdered binder. To increase the charge magnitude and the content of homocharged particles, triboelectrification can be combined with other charging techniques, for example, with the charging in the field of corona discharge of the same polarity [38].

Schemes of apparatus for unilateral and bilateral prepregs formation by electrodeposition are shown in Figure 7.12 and Figure 7.13, correspondingly. Table 7.7 represents triboengineering characteristics of the composites based on polycaproamide powder.

According to the data from Table 7.7, EIT provides better triboengineering characteristics for all parameters than the conventional scattering. This is explained by the tighter deposition of the binder layer at electrostatic combination of the components and also by the polymer molecular structure orientation in the field of the binder's residual charge during prepreg manufacturing. As a result, the polymer matrix obtained has fewer defects and more uniform distribution in the filler framework. Considering the possibility of a wide-range variation of the material composition and its environmentally-friendly manufacturing, the EIT can be described as promising technology for the composites production. Undoubtedly, in the majority of cases the electrization and the electrostatic effects must be taken into account manufacturing of the woven composites [9,39–43].

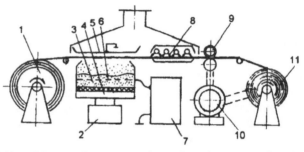

Figure 7.12. Scheme of apparatus for unilateral prepregs formation in the electric field: 1, drum with the filler band; 2, vibrator; 3, pseudodeposition chamber; 4, air supercharge chamber; 5, screen; 6, electrode; 7, high voltage source; 8, thermochamber; 9, clasping rollers; 10 electromotor; 11, drum with the prepreg.

Table 7.7. Triboengineering properties of compositions

Filler	Method of the prepreg formation	Specific load, MPa	Wear rate, 1×10^{-8}	Friction coefficient	Temperature in the friction zone, K
Carbon fabric	Electron-ion technology	5	4.81	0.05	346
	Scattering	5	9.84	0.12	375
Fiberglass fabric	Electron-ion technology	1	9.84	0.23	344
	Scattering	1	17.01	0.50	380
Sliding speed $v = 0.5$ m/c					

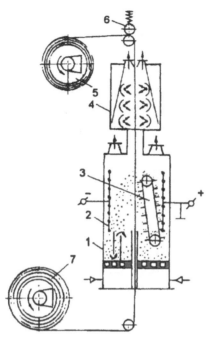

Figure 7.13. Scheme of apparatus for bilateral electrodeposition of binder on the filler band: 1, pseudoliquefying device; 2, electrodes; 3, conveyer; 4, thermochamber; 5, drum with the prepreg; 6, clasping rollers; 7, drum with the filler band.

7.6. PROSPECTIVE APPLICATIONS OF TRIBOELECTRIC PHENOMENA

Local mechanoemission effects in the friction zone of dielectrics are accompanied by the electron-ion bombing [44]. Bombed surfaces lose their physico-mechanical properties - microhardness, elastic modules, and so forth, become lower. For example, the elastic modules of copper foil and duralumin decreases by 10 to 60 % after such electron-ion bombing. The contact of dielectric materials and the artificial electrization of mineral particles for sand-blasting or grinding increase the productivity by 1.5 to 4 times.

Electron-ion processes accompanying the external friction cause the twofold reversible changes in microhardness, strength, electrophysical properties, adsorbability, and crystal lattice imperfections of the rubbing surface layers. Controlling this phenomenon assuming artificial electro-ion bombardment (EIB) can lead to realization of the controlled technological processes. Experiments have shown that at friction and by machining controllable changes of plastic, hardness and elastic properties of the solids are possible by applying the EIB. In particular, the EIB increases wear resistance of polymers and rubber, is twice as fact as sand-blasting and grinding processes, decreases the ground surface roughness, and increases by 2 to 10 times the grinding wheel durability.

It should be mentioned that the charging and the changes in conductivity at machining have been reported as early as the last century by Faradey. It was later established that the charge transferring at machining takes place due to small distances (~2.5 Å) between the contacting bodies. Charge transfer in dielectrics is realized due to the existence of the electron surface states with quasizone structure. Machining affects the formation and the interaction of the excited states resulting in elevated temperatures, luminescence, electron emission, appearance of surface charges, and so forth. These phenomenon can be neutralized (at least partially) by passing the electric current, using special technological lubricants, and so on.

Medicine is one of the unusual fields of application of the polarization. In 1983, Gromov and Krylov have determined laws of dissolution of the polarized solids. They showed that the dissolving rate of solid dielectrics (ion crystals, glasses, organic compounds, et cetera) polarized prior to or during the dissolving process changes with the polarization degree. If dielectric permeability ε of the dissolved substance is higher than that one of solvent, the molecules are "retained" by the solid body (slowing down the dissolution) and vice versa [45].

There are reasons to believe that complex electrophysical and electrochemical processes take place in the biological friction assemblies – human joints [46,47]. Yet, this problem should be thoroughly studied.

The list of insufficiently explored tribophysical problems having direct practical applications is quite long. It is obvious that active triboelectric phenomena take place at modification of the rubber bushings by ion implantation [48], laser treatment of polymer surfaces [49], coating of elastomers with PTFE by electrophoresis [50], vacuum metallization of polymers [51], and other high-energy processes.

Mechanoemission can detect the destruction of adhesive and cohesive bonds in solids. By measuring the mechanoemission rate at vibrodispersion, one can estimate the possibility of use of particular mineral as an active filler in organomineral materials made by mechanoinitiated polymerization [52]. Mechanoemission can be also used in radiation polymerization and radiation-chemical modification of the solid surfaces. Moreover, mechanoemission could be used in tribodiagnostics. For example, continuous control of the friction process can be based on the work function measurements [53].

Considering the contact electrization, it was proposed to use triboelectromotive force and its quasistationary model as a method of a friction pair studying and diagnostics [54]. Broad prospects of triboelectric research exist in connection with modern office technology. The limits of their frictional parts durability have not been reached yet.

The data presented do not completely reflect all applications of the mechanoemission, triboelectrification, and the electret effect. They are numerous and have, far-reaching implications..

CONCLUSION

For several decades tribologists observed and tried to explain numerous emission effects resulting from tribodeformation of the material surface layers. Electron emission, electromagnetic radiation (from radio to X-ray), and different acoustic and low-frequency oscillations can render a significant amount of information. However, their registration and interpretation are quite difficult since they appear directly at friction interaction only, which results in difficult to measure morphological and structural surface changes. It is possible that the surface electrization remains the only universal and easy-to-identify witness of the macro- and microscopic "dramas" at friction.

The understanding of triboelectrification mechanisms started with developments of quantum-mechanical models in surface physics and chemistry, ideas of the electron energy levels, and molecular forces theory based on ideas of fluctuating electromagnetic fields in condensed materials. Polymer materials have appeared to be highly sensitive to tribolelectric phenomena. Their studies have significantly expanded the knowledge of the structural mechanisms of tribodeformation and

186   APPLICATION OF THE EFFECTS

tribodestruction of solids. Quite important was the disclosure of triboelectret effect resulting from the triboelectrification of polymers. Interrelation revealed between the triboelectrification and the triboelectretization has enabled further developments in the friction and wear theory and has identified the fields of their practical application.

Electric and electromagnetic phenomena have been observed and studied for hundreds years. However, the appearance of new materials and technologies has caused a new wave of interest in these phenomena. New theories and applications for triboelectric phenomena are still to be discovered.

REFERENCES

1. Bilik, Sh.M. 'Metal-Plastic Friction Pairs in Machinery' (in Russian), Moscow, 1965.
2. Bilik, Sh.M., Tsurkan, V.P. //Application of polymer-based materials in machine friction pairs and sealings (in Russian), Moscow, 1968, 46–48.
3. Distler, G.I. and Moskvin, V.V. // Reports of the USSR Academy of Sciences, 1971, 201, no.4, 811–893
4. Malozemova, T.I., Klimovich, A.A. and Rutto, R.A. in 'Scientific and Engineering Achievements in the Field of Filled Polymers Used in Machinery' (in Russian), Moscow, 1987, p.48
5. Lebedev, L.A. and Georgievskii, T.A. in 'Electrochemical Processes at Friction and Their Use in the Fight against Wear' (in Russian), Odessa, 1973, p. 15
6. Sviridenok, A.I., Belyi, V.A., Smurugov, V.A. and Savkin, V.G. Wear, 1973, 25, 301–308
7. Briscoe, B.J. ACS Symp. Ser. 1985, 287, 151–170
8. Bahadur, S. and Tabor, D. // Wear, 1984, 98 1–13
9. Sviridenok, A.I. in 'Tribology in the USA and the Former Soviet Union: Studies and Applications', Allerton Press, N.Y., 1994, p.157
10. Sviridenok, A.I. and Meshkov, V.V. in 'Tribology – Solving Friction and Wear Problems', Technische Academie, Esslingen, 1996, 3, p.2347
11. Evdokimov, Yu.A. and Kolesnikov, V.N. in 'Friction, Wear and Lubricants' Vol.2 (in Rusian), Moscow, 1985, p.308
12. Evdokimov, Yu.A. and Kolesnikov, V.N. Journal of Friction and Wear, 1993, 14, no.2, 289–307
13. Vladykina, T.N., Toporov, Yu.P. and Luchnikov . Journal of Friction and Wear, 1988, 9, no. 3, 534–538
14. Gan, G.K. Journal of Friction and Wear, 1991, 12, no. 1, 81–88
15. Garkunov, D.N. in 'Fundamentals of Tribology' (in Russian), Moscow, 1995, p.225
16. Shpenkov, G.P. 'Physicochemistry of Friction' (in Russian), Minsk, 1991
17. Larsen-Basse, J. in 'Korea – US Tribology Symposium', Seoul, 1995, 1–7
18. Vasilenok, Yu.I. 'Prevention of Static Electrization of Polymers' (in Russian), Leningrad, 1981

19. Bliznets, M.M., Zaitsev, A.L., Bogdanovich, P.N.and Dyatko, E.K. 'Wear-proof
     Composites Based on Thermosets' (in Russian), Minsk, 1987
20. 'Polymers in Friction Assemblies' Ed. by A.V.Chichinadze (in Russian), Moscow,
     1988
21. Pinchuk, L.S., Struk, V.A., Myshkin, N.K. and Sviridenok, A.I. 'The Material
     Science and Engineering Materials' (in Russian), Minsk, 1989
22. Sviridenok, A.I., Meshkov, V.V. and Shuvalov V.B. Exploitation Problems of
     Machines, 1994, 29, Warsawa, p.575
23. Voronezhtsev, Yu.I., , Pinchuk, L.S. and Snezhkov, V.V. 'Electric and Magnetic
     Fields in the Polymer Composite Technology' (in Russian), Minsk, 1990
24. Kestelman, V.N. Advantages in Polymer Technology, 1995, 14, no.3, 243–247
25. Mironov, V.S., Voronezhtsev, Yu.I., Klimovich, A.F. and Goldade, V.A. in
     'Friction, Wear and Lubricants' Vol.5 (in Rusian), Tashkent, 1985, p.50
26. Bocharov, A.M., Sysoev, P.V., Mironov, V.S. and Klimovich, A.F. Diamonds and
     Superhard Materials (in Rusian), 1983, issue 3, 1–2
27. Mironov, V.S., Bocharov, A.M. and Klimovich, A.F. Superhard Materials (in
     Rusian), 1985, issue 3, 61–64
28. Epifanov, V.I., Pesina, A.Ya. and Zykov L.V. 'Technology of Diamond Processing
     into Brilliants' (in Russian), Moscow, 1982
29. Bocharov, A.M., Sysoev, P.V., Mironov, V.S. and Klimovich, A.F. Diamonds and
     Superhard Materials (in Rusian), 1983, issue 3, 1–2
30. Klement'ev, N.M. 'Thermodynamics of Friction' (in Russian), Voronezh, 1971
31. Korobov, Yu.M. and Preis, G.A. 'Electromechanical Wear at Friction and
     Machining of Metals' (in Russian), Kiev, 1976
32. Dovgyalo, D.A. and Yurkevich, O.R. 'Composite materials and coatings Based on
     Dispersed Polymers' (in Russian), Minsk, 1992
33. Klimovich, A.F. 'Creation and Study of Electrostatic Technique of Thin-layer
     Polymer Coatings Manufacturing' Dissertation (in Russian), Gomel, 1970
34. Krasovskii, A.M. and Tolstopyatov, E.M. 'Manufacturing of Thin Films by Polymer
     Vacuum Scattering' (in Russian), Minsk, 1989
35. Shustov, V.P., Sviridenok, A.I., Sikanevich, A.V. and Gaiduk, V.F. in 'Resourse-
     Saving and Environment-Friendly Technologies' Part 2 (in Russian), Grodno, 1995,
     p.217
36. Gromyko, Yu.V. and Kravtsov, A.G. Reports of the Belarus Academy of Sciences
     (in Russian), 39, no.5, 112–116
37. Markov, E.M., Pinchuk, L.S., Goldade, V.A., Gromyko, Yu.V. and Choi, U.S.
     Journal of Friction and Wear, 1995, 16, no. 3, 518–522
38. Mironov, V.S., Dovgyalo, D.A. and Yurkevich, O.R. Journal of Friction and Wear,
     1991, 12, no.2, 326–332
39. 'Textile Structural Composites' Ed. by Tsu-wei Chou and Frank K. Ko, Comp.
     Mater., ser.3, Elsevier, Amsterdam, 1989
40. Friedrich, K. (ed.) 'Friction and Wear of Composites', Elsevier, Amsterdam, 1986
41. Ponomarenko, A.T., Shevchenko, V.G., Kryazhev, Y.G. and Kestelman, V.N. // Int.
     J. Polymeric Mater., 1994, 25, 207–226
42. Kestelman, V.N. and Evdokimov, Ju.M. Plaste und Kautschuk, 1992, 10 345–346
43. Eiss, N.S. in 'Tribology in the USA and the Former Soviet Union: Studies and
     Applications', Allerton Press, 1994, p.183; Pogosian, A.K. 'Tribology in the USA

and the Former Soviet Union: Studies and Applications', Allerton Press, N.Y., 1994, p.271

44. Balabekov, M.T., Balabekov, Sh.T., Usmanov, Yu.M et al. in 'Friction, Wear and Lubricants' Vol.2 (in Russian), Moscow, 1985, p.330
45. *Discoveries in the USSR* (in Russian), Moscow, 1985
46. Kupchinov, B.I., Rodnenkov, V.G. and Ermakov, S.F. 'Introduction in Tribology of Liquid Crystals' (in Russian), Gomel, 1993
47. Furey, M.J. and Burkhardt in 'Tribology – Solving Friction and Wear Problems', Technische Academie, Esslingen, 1996, 3, p.1421
48. Guseva, M.N., Lysenkov, P.M., Sokov, E.V. and Vladimirov . *Journal of Friction and Wear*, 1993, 14, no.4, 742–748
49. Nikitin, L.N. and Said-Galiev, E.I. *Journal of Friction and Wear*, 1994, 15, no.1, 149–165
50. Kuzharov, A.S. and Danyushina, G.A. *Journal of Friction and Wear*, 1990, 11, no.2, 287–294
51. Lipin, Yu.V., Rogachev, A.V., Sidorskii, S.S. and Kharitonov, V.V. 'Technology of Vacuum Metallization of Polymer Materials' (in Russian), Gomel, 1994
52. Toporov, Yu.P. in 'Materials for Radioelectronics' (in Russian), Moscow, 1985, p. 25
53. Zharin, A.L. *Journal of Friction and Wear*, 1993, 14, no.3, 570–583
54. Vasil'ev S.V. *Journal of Friction and Wear*, 1990, 11, no.3, 526–531.

ELECTROPHYSICAL PHENOMENA IN THE TRIBOLOGY OF POLYMERS

A.I. Sviridenok, A.F. Klimovich, and V.N Kestelman

SUBJECT INDEX

elastic 20, 26
metal-dielectric 47
metal-polymer11, 18, 29, 30, 36, 37, 52, 87, 92, 111, 134, 157
plastic 83
static 23
Contact potential difference 8, 23, 24, 100
Current inversion 69, 149, 158
Curves of potential distribution over the surface 123

D
Decoration of charged surfaces 25, 37, 54-58, 119, 120
Defects of structure 3, 15, 17, 126, 133, 177
Dielectric permeability 7, 8, 10, 22, 42, 71, 76, 82, 84, 135, 183, 184
Discharge 3-5, 10, 18, 20, 22, 29, 38, 55, 85, 94, 96, 116, 121, 124, 129, 130,
144-146, 149, 166, 166, 174, 176-180
Double electrical layer 6-10, 15, 17, 85, 96, 141

E
Effective surface charge density (ESCD) 21, 41, 43, 47, 50, 102, 105, 106, 108, 116-119, 121, 122, 127-133, 135, 137-139, 149, 150, 152, 153, 167, 173
Electret state 3, 4, 14, 28-31, 41, 47, 50, 51, 83, 85, 87, 95, 97, 98, 101-106, 111, 113, 116, 117, 144, 146, 147, 149, 151, 154, 162, 167, 178, 179, 181
Electrets 18, 20, 21, 25, 27-31, 41, 53, 54, 87, 94, 95, 98, 100, 105, 106-109, 144-146, 152
Electrical relief of surface 53, 56
Electrification
contact 4, 15, 23, 24, 62, 121
of melts 100
of polymers 30, 37, 38, 41, 49, 74, 83, 85, 100, 141, 146, 156, 157
static 22, 25, 29, 46, 47, 61, 87, 152
in vacuum 50, 156
Electrification current 36, 84, 87-89, 95, 97-100, 102-107, 116, 148, 149, 157-160
Electrophysical phenomena 4-6, 9, 14, 21, 22, 118, 152, 155, 160, 168
Electrostatic interaction 111
Emission 3-6, 9, 10, 20, 22, 36, 47, 49, 119, 123, 124, 134, 144, 145, 156, 157, 184

T - #0192 - 101024 - C0 - 229/152/11 [13] - CB - 9789056995775 - Gloss Lamination